THE LOGICAL SOLUTION
OF THE
RIDDLES OF GRAVITY
AND
EINSTEIN'S THEORY

THE LOGICAL SOLUTION
OF THE
RIDDLES OF GRAVITY
AND
EINSTEIN'S THEORY

George Kirakosyan

To order additional copies of this book, contact:
Xlibris Corporation
1-888-795-4274
www.Xlibris.com
Orders@Xlibris.com
33659

CONTENTS

SUMMARY

In the offered work, unlike the majority of formal theories, the author considers a problem of gravitation with causally consecutive point of view. By logic of researches of known facts and existing theories of this direction, he comes to conclusion about unknown fundamental property of a matter causing the gravitational phenomena. The author's explanation is not difficult to mastering both with causal and quantitative points of view. However, it demands deep changes in accepted beliefs and approaches. On the basis of the offered causal essence of gravitation, the author deduces *the law of universal gravitation of Newton*; he defines the theoretical value of a *gravitational constant* and calculates known *gravitational effects*, with the use of simple mathematical reasoning only. The subsequent results of the offered concept are corresponding with the checked-up results of Einstein's *general theory of relativity (GTR)*. However, for planned new experiments, on detection of *gravitational waves* and *gravymagnetic effect*, negative results are predicted. The mentioned circumstance may allow judging about value of the offered explanation in a not-far future.

Despite of complexities of the studied problem, the book is narrated in a free, polemical style, stipulated for a wide range of readers.

---

*   The picture used on the cover is the *Galaxy Andromeda M 31*, from site www. ASTROLAB.ru

INSTEAD OF PREFACE

Usually, not all forewords become carefully studied, especially, when those are long. On it, I shall briefly note only for whom this book may be stipulated and how to study it. Previously, we shall conditionally divide the people on two main categories without intention to humiliate any of them. The political, financial, sports and other events connected with human activity often much more are important than others for the many of individuals. This book, probably, will be out of their interests, in author's view. There is also the type of a rare people for whom it is more interesting why the stars are blinking in the sky than, let's say, life's details for any politician or famous showman. The studied matter is about events and phenomena of nature, which are not transient together with human history but exist forever. Hence, the book will be interesting mostly for the second type of people, who is ready also to certain efforts of satisfying their own inquisitiveness. How to study it easier? It's depending on many conditions—from the previous acquaintance of a reader with the studied problems, from his own patience, etc. To that purpose, I can suggest only one simple way tested by me. That demands to carefully read the book several times with some intervals, temporarily forgetting about arising questions. With time, the reader will have felt that the previous number of questions continually decreases and depicted image becomes more complete. I hope, after natural doubts and serious judgments, he will become confident that presented explanations are something more than only a witty story.

INTRODUCTION

"But he has nothing on at all," said a little child.

Andersen—"The Emperor's New Suite"

The old mystery of gravitation within long years has occupied the imagination of scientific experts and many inquisitive people far standing from the theoretical physics. To estimate the cognitive value of that problem in science, it would be enough to pay attention to absolute ignorance till these days, of causal explanations for such phenomena as the weight, felt by us constantly, and to the free falling, meeting everywhere, despite the quantity of written pages and spent efforts on this direction. Since the times of Great *Newton*, the all-achieved success of theorists on studying the gravitational phenomena, actually, concern only the quantitative side of its description and didn't yet bring to appreciable advance on a cognitive direction. Some hypotheses had been suggested through time toward explanation of the physical essence of gravitation, such as suppositions about of particles: *Lesagions, Gravitons,* etc. However, all of the similar theories remain unsatisfactory until now, in the sense, as the ways to deduce the *Newton's law of gravitation* as well as to define the value of *gravitation constant* aren't given into them on a conceptual basis. The heuristic meaning of the *Einstein's general theory of relativity (GTR)* and of other alternative theories of gravitation may be estimated also as nonsatisfied from cognitive point of view, as the marked problems aren't solved in them either. In the created situation, the suggestion of new book about gravitation, most likely, may be perceived with lawful skepticism as another small useful exercise in the area not clear for anybody; such examples are much. In developed reality, to gain trust and interest of readers, the author cannot think anything better than a preliminary representation of the purpose of his book and its difference from already-existing numerous works concerning the problem. From this point of view, the writer considers correct to notify the reader about his own suggestion of the explanation of the gravitational phenomena from beginning, without hypocrisy or crafty modesty. In his opinion, its clear formulation enables to represent it in an accessible form, stipulated for a wide range of public, despite of difficulty of the theme. The author has accepted such decision, concerning the style and to the destination of the book, after serious doubts conditioned by a bureaucracy in official science, and from other certain circumstances, which will be necessary to discuss in details. First of all, it was taken into consideration that debatable areas of physics, as a rule, are divided into separate schools and directions; each of them is considering its own principles and outlooks, sometimes significantly distinguishable from each other. Besides, due to its technological value, proceeding from national, patriotic state, politically or from other egoistic reasons, it is accepted to keep the problematic, advanced edge of science under lock in general, thus, inaccessible to a society. The corresponding rigid system of rules and an official caste of scientific functionaries are formed now on the described ground. The paramount purpose of that is to keep science under the supervision rather than to contribute to development or to unforeseen spreading of it. There is nothing too surprising. It's well-known now that knowledge isn't an innocent leisure only, but it's also technology, wealth, and authority. Hence, the desire to hold it under supervision becomes clear. Because of described and other explainable circumstances, the fundamental science, being an open and attractive part of philosophical, natural sciences at the beginning, presently become the one important sphere of a policy and business. The some purposeful reserves and rules are introduced by the mentioned caste and successfully apply that now to the delicate protection of the borders of a privatized scientific possession and to preserving it from extraneous encroachments or from other attempts of revolutionary change. From this point of view the *Copenhagen's interpretation* has served it best, accepted in physics in the beginning of the past century. As per the named new ideology, the *quantum or statistical laws of the nature* are considered to be in the basis of existing material world, instead of a *cause-consequential laws*, accepted earlier "by mistake." Because of the new ideology, the *formal mathematical methodology of research* has begun to be applied in physics in place of the old *cause-logical* way of the analysis. Subsequently, the physical science has gradually lost its former attractive clearness and heeled aside of abstraction, thus becoming an elite discipline, torn off from its philosophical root as well as from the society. However, after transient successes of *quantum theory*, some innocent, "promising new theories," similar to a *quarks theory* (still promising for many decades), are issuing from time to time to justify the subsidized huge means and to creations of visible progress, in many respects, in the already-fruitless science. There are also many next "sensational openings," by an example of the new kind of deviation, revealed in traces of the accelerated particles after their colliding, in addition to the already-known thousands, etc. At the same time, the idea of improbable complexity of fundamental problems of physics and inability of nonspecialists to penetrate into them is assiduously inspired by the inhabitants, thus justifying the access of elected people only to that important part of universal cultural treasure. These words not necessary to interpret unequivocally that all categories of people, called "experts of physicists," are definitely conservatives or they are deliberate malefactors. The many fair persons and gifted minds among them in reality. Rather frequently, many of them guess also that physics does not go toward its designation. However, the policy and morals in this area are combined in such a manner that an oppositionist remains either to obey to a "duty of service" without resistance, or, under pity or sneer of more prudent colleagues, to leave his familiar environment, sometimes the loved occupation too. Adding to the already-mentioned circumstances such humanly clear reasons as the social career interests and ambitions of ruling persons in science, the reader can judge about the possible reception in the official environment of an unknown author's new approach, demanding radical changes in the established beliefs and involuntarily touching to the already-recognized, well-known authoritative opinions too. The similar prospect and described reality have induced the author to design his work not for the functionaries or officials, but to destine it to the inquisitive part of scientists and society, with hope on the court of time. At the same time, the author definitely is not excepting an opportunity to display an interest toward the book from the side of firsts as well, as any more or less significant achievements anyhow becomes subject to business and policy, whether we desire that or not.

Except imagining approximately the developed conditions in present fundamental science and around it, another important point for estimation of the book and to its successful mastering is the primary confidence of the reader in his own abilities and opportunity to talk about fundamental scientific problems in easily understandable and accessible language. For overcoming the uncertainty and the implanted prejudices in this direction, we shall now look at some simple reasons and explanations, connected with the issue interesting to us. As it has been already mentioned above, the existing theories about gravitation, only its quantitative side had been studied and established only. In other words:

The modern physical theories are the formal mathematical constructions that are putting the problem of definition of kinds of mathematical formulas and equations most precisely expressing the quantitative parities of investigated phenomena without mentioning about their causally consequent sides.

A little bit different way of solving the tasks is applied in the offered book—by starting it from disclosure of the causal mechanism of occurring phenomena. Such divergence of methodology is caused by the above-mentioned deviation of the physical science aside from mathematical formalism in general. The basic difference between described methods is possible to be characterized by the following words: In the first case

The schemes or recipes are searching or developing by the corresponding ways, on a task of adapting the mathematics to a description of established facts, the causal sides of which remain unexplored.

Meanwhile,

The author primarily is deducing the causal essence of studied phenomena by logical investigations, from which the required quantitative parities follow naturally, describing and confirming the fidelity of initial decision.

There is nothing unusual or fantastically new for anyone in the above statement. It simply means returning to logical analysis, natural and habitual to us, from which the modern physics are far considerably. Moreover, in the modern methodology on difficult situations, it allows the use of new hypothetical essences and uncertain concepts also, in unlimited quantity. There, too, allows the special adjustment of the mathematical formulas and parities on need, for example, by introducing of some "calibrated," "rated," or other functions. Thus, in modern physics, actually, the composing is authorized, depending from taste and conscience of researchers and from approval of its main inspectors. Because listed and other logically defects of accepted methodology in a modern physics, thorough disagreements have risen between philosophers and physicists now. The leaders of physics oppose fair conclusions and criticism of firsts their little bit original arguments.

They declare, for example, that the problems of physics should not concern the philosophers in general, as with their "shabby science" they didn't yet create any such valuable thing for humanity, which may be comparable with the merits of physicists! Considering such state of affairs, the reader himself can judge why we should feel defective or confused when we don't understand the overloaded, formal language with uncountable, conditional names, composed by modern theorists, used in their diligences to explain something, which continues to remain dark for us and for them too. We have repeatedly found out the opposite instructions in history of natural sciences. That is, as soon as the natural science become overloaded, formal and little productively, as a rule, it was evidencing about its significant divergence from correct and rational way, which was demanding radical revolution both in the concepts and in the methods of its development. The similar reasons show to us which way to prefer. The presented explanation opens up the base of difference between author's approach and the formal methodology used nowadays. That shows unnecessary to detailed studying of all prior, existing huge volume of complex mathematical reasoning, abounding in modern formal theories of this direction, demanding great efforts and time. For correct orientation in such circumstances is enough to remember only one fact: *none of the formal theories has led to a certain success to disclosing the causal essence of gravitation.* Such circumstance allows everyone to state his point of view and to offer his own decision of the riddle in a form he favors, without moral damage or infringement of authoritativeness of anybody, in author's opinion. That will be incorrect to interpret the told as deletion or humiliation of a created and already-achieved results, on which, certainly, nobody has the right. The author is deeply grateful and esteems the founders of many works over the subject interesting to us, existing today. Proceeding from these, actually, it has become possible to approach the understanding of the essence of gravitation, by selecting all the rational from them, then criticizing and rejecting all unnecessary, which has always been an allowed approach on the way of development of natural sciences. However, the scientific objectivity and cult to authoritativeness are deeply incompatible things. The author is compelled to remind this banal truth to those who, in noble impulse of soul, will hasten to see an attempt to "sacrilege" to the deserved names in physical science, to a cheap sensation, or to other low motives in below-mentioned critical remarks. Meanwhile, the true reason of occurrence of these lines is very simple. Deeply convinced in correctness of his conclusions the writer with natural desire has decided to share that with those to whom it will represent any interest.

SOME IMPORTANT CONCLUSIONS FROM GTR

Sorry Mr Einstein. I can't find the "space-time" in your theories.

Under pressure of new problems in history of natural sciences, it has already happened many times, to doubt in the fidelity of long-established beliefs accepted as far from any suspicion and to refuse that resolutely. The vain attempts of many years to find out the physical essence of gravitation, simply, demand from us to reconsider, with objective care, both the analytical methods, used nowadays, and all the initial representations, connected to the discussed issue, despite their authoritativeness or on a quantity of their supporters. Those who tried to understand anything distinct about the physical essence of gravitation, for example, from formal theory of GTR and the numerous popular literature accompanying it, probably, may remember a number of unusual names and word combinations by means of which the origin of the gravitational phenomena are explained in it. The statements about "indivisibility of coherence of space-time," about its "curvature," etc., may deserve as the examples told. In offering a new explanation of gravitation now, the lawful question about its mutual relation to the mentioned concepts and, in general, to already-existing theories of gravitation will be primarily interesting for the reader. To illustrate this important question, it is necessary to take into consideration that such impressing names and expressions are applied in formal theories for designations of the kinds of quantitative parities, or ways of mathematical actions, etc. Thus, these names have first of all the relative mathematical meanings, for example, for designation of some type of system of equations and so on, they are not stipulated to express somewhat of concrete-physical sense. Thereby, the different speculations with such names should be considered inexcusable, which frequently find places in the works of formal sense researchers and of scientific visionaries. For example, such conclusion from the fact of "curvature of space-time" can be met in the literature—As it is curved, it is possible to search an opportunity to move in that by shortest way, thus, reaching to a completely unusual results! Generally speaking, the detailed analysis of meaning or values of all concepts, used in previous works, devoted to the discussed question, should not be our duty, as for each new theory, only the condition of its conformity with the existing facts can be considered, and its compliance with the already-existing representations or terminology is not obligatory. However, some preliminary explanations in this field are necessary, as the burden of a half-truth and mysticism involuntarily presses on our consciousness up during many years and sometimes even prevents us to trust the elementary logic. To understand its essence and to truly evaluate the formal methodology, it is better to recollect our school difficulties, which we all had while doing our homework. To find out the required answer, without overloading our "brain device," we were frequently trying to get it by multiplying, dividing, or other operations with the initial data of the problem. After getting the answer in such a way, which was sometimes possible, a more or less reasonable explanation to maintain our mathematical actions was composed to satisfy our class teacher, which we were not always succeeding to do. A similar approach, without special exaggeration, is applied in modern formal theories also.

Without involving the reader into nonperspective, logical discussion of methods and basic statements, used to construct the GTR, we will only consider a few clear notions connected to it.

As we know, Great Newton does not give any answer to a concrete question: "How does the gravitation originate?" by confessing fairly that he knows nothing about it. The explanation of this phenomenon is ostensibly given in GTR, which in brief sounds approximately as follows:

Under the influence of material body, the space-time in its environment becomes "distorted." As a result of influencing on the various bodies, it's forcing to move them with acceleration toward the central gravitation body and to press on its surface, creating the known weight's forces.

Without discussing now, the natural questions rise from such explanation: what concrete-physical should be meant under the name "space-time"? How is it "distorted"? Moreover, how the "curved space-time" influences matter, etc., we shall only notice the following: to explain the observable phenomenon, to the existing and totally concrete two subjects (central and trial bodies) a third name, and a new participant in the phenomenon is added here. That is the "space-time," the independent existence of which is impossible to prove. Nevertheless, we already know from historical experience that similar practices—the reference toward the hypothetical new participants of phenomena in difficulty cases—already and repeatedly have taken place in physics, which have brought nothing useful to science but new mysteries. The episodes with the "Phlogiston," with the "weightless electric liquid," with the "world ether," etc., invented to explain experimentally defined incomprehensible facts can serve as instructive examples to this statement. The realistic thinkers have never respected the described approach (it's better to say—the method of composition). To this case, we can recollect the known statement of Newton—*I am not building the hypotheses.* It has long been criticized by philosophers (see the *Ockam's principle* or *the Ockam's razor*) and at last by Einstein himself too, demanding to operate with concepts in scientific theories the necessity of which is undoubtedly dictated by the experiments and by measures (see the *operationalism of Einstein)* [L-2]. Based on what's told, it is rather difficult for us to explain or to estimate the occurrence in Einstein's theories of a "space-time," intangible and undetectable but quite responsible for completely real consequences!

It is possible to get the true estimation of the meaning of the "space-time" by close study of the quantitative sites of Einstein's theories. Hereunder, the following words can have rather-strange impression, but they only ascertain the existing fact. The mater is in that:

The somewhat new physical constant that could be considered as the quantitative characteristic of the "space-time" does not exist in formulas of STR (the special theory of a relativity) as well as in the GTR about which is spoken in them.

Thus, if somewhat of reality participates in the phenomenon, then the corresponding constants should be presented in quantitative descriptions of that phenomenon, which individually characterize that hypothetical participant. For example, such constants for the electron are the values of his mass and the electrical charge, shown in the equations, describing laws of radiation of the atom and showing that it really participates in this process. The absence of the new physical constants in Einstein's theories with all obviousness shows that the "space-time" has only the verbal, psychological value in them. Thus, it is as painful as it will be to realizing that the many years of discussions and speculations around the "space-time" and about its features were losing the time and have only withdrawn us away from the true way of thinking.

Many kinds of arguments were examined and uncountable pages were written about the "space-time," but until today, except of a formal mathematical interpretation, nobody knows resolutely what concrete-physical's image is under this name. Meanwhile, from logical point of view, GTR contains also very valuable instruction, which can guide us toward disclosure of the true essence of gravitation. This concerns Einstein's another mysterious statement—toward *"local equivalence of gravitation and inertia"* accepted as one of the main principles for construction of GTR. It is necessary to emphasize specially that unlike "curved space-time" the mentioned statement has its concrete, quantitative expression in GTR consisted in the acceptance of *equivalence of the gravitational and inert mass.* Not to disturb the reader with the above used terminology, now we shall tell only the next:

The fact of falling of all trial bodies with identical acceleration independent of structure of their materials, forms, or sizes opened by Galileo for the first time is accepted in GTR.

The falling of various bodies with identical acceleration seems rather surprising from cognitive point of view and in many respects reflects the mysterious essence of gravitation. In GTR, this quantitative parity is accepted as experimentally established fact, without weighty causal interpretation, covering such ignorance by the "curved space-time." To the benefit of fairness of these words, it is possible to refer to Einstein's known words: "If I would only understand, what actually occurs in the cabin of falling lift!" These words tell us evidently that Einstein initially did not put any definitely cognitive, physical, or causal sense in the concept of "curved space-time." Based on such explanations:

We should refuse the usage of "space-time" if we have sufficient boldness to admit its vague physical sense and if we really have an intention to search the explanation of gravitation.

It is also very important to mention that another important initial condition, which has its quantitative expression as well, is accepted in GTR. Actually, it is a *consideration in GTR the limitation of speed in measuring*, which we shall speak in details a little later. Thus, leaving aside some verbal statements of GTR, concerning the "space-time," and considering only its rational contents, from cognitive point of view, it is possible to characterize it by the following words:

The GTR leads to a certain experimentally confirmed results by accepting the equivalence of gravitational and inertial phenomena (locally) and the finiteness of the speed of interactions in measuring.

By further studying the quantitative reasons of GTR, it is possible to conclude that in Einstein's accepted principle—"local equivalence of gravitation and inertia," the restriction of "localness," except of verbal—declaratively, actually, it doesn't have any quantitative expression or an experimental substantiation itself. A Canadian physicist *Paul Marmet* has noticed it, for example, as well. After studying the book, it will be clear for the reader that the mentioned restriction of "localness" has appeared under pressure of intuitive perception of reality only, hindering the comprehension of the true essence of gravitational phenomena.

What's told above definitely shows that the suggested explanation of gravitation doesn't mean fully rejection or deletion of the GTR and other less-known theories of gravity.

The GTR and other alternative formal theories are looked from cause-consequential point of view with the cause-logical comprehension of gravitation. Thus, they become clear and free from their unnecessary parts, in author's confidence.

ACQUAINTANCE WITH
USED CONCEPTS

There is version according to which Alexander the Great has undertaken the
military campaigns to be reaching to the Earth's edge . . .

Above, it has been mentioned the possibility to the comprehension of the
nature of gravitation by way of critical discussion of main principles of GTR and
by logical analysis of some experimentally established facts that will be shown in
the following pages.

*The all-gravitation effects known from GTR, the theoretical conclusion of Newton's
law of gravity, and the definition of gravitation constant become possible to
represent in causal and quantitative interpretations on the basis of offered
explanation of gravitational phenomenon.*

The marked results are pointing to rightness of the offered explanation by
sufficient confidence. However, for prejudiced skeptics, an opportunity is present
always to attribute the mentioned decisions to rather-surprising coincidences.
Such apprehension is probable, and to overcome it is one of the most important
problems that we are facing. The suggested explanation of gravitation with its
oddness deeply contradicts the intuitive perception of the surrounded material
world and demands thorough revision in our imaginations and in existing beliefs,
which has always been a psychologically difficult problem. The sense of what's
necessary to tell about physical essence of gravitation may be formulated briefly
in several sentences. However, a significant preparatory work is required for its
perception, which carries mostly psychological rather than educational character.

If to follow all of reasoning toward disclosure of the gravitation's essence in a
sequence by which they have been made, it will be a long and rigid job for the
reader. Besides, this way demands some acquaintance both with Einstein's
mentioned theories (STR and GTR) and with several critical works, too,

concerning them, which will be unfair toward our imagined reader. Issuing from such reasons and aiming possible easiness of the suggested material, the author has made another decision. We shall begin with discussions of methodological principles to analyses and some well-known established facts, for a while leaving aside Einstein's theories and others alternative theories about gravitation as well. We will do it without aspiring to strict and unconditional substantiation of all statements and conclusions, which is simply not possible in the stipulated volume of this book. However, we shall come back to existing theories from time to time, for parallel comparisons and for new interpretations of established results, as the question of gravitation anyhow cannot be separated from the Newton's law of gravitation and from the mentioned works as well. It is possible to find out some useful examples from instructive experience of history of the natural sciences and make some conclusions that may help us to decide from what side to study the problem of gravitation. We can notice, for example, that each time when a deadlock situation has arisen in the science, as a rule, the solutions has been found by thorough revision of several initial representations and by refusal from false beliefs considering before as indisputable.

The past beliefs in the opportunity to define the *absolute rest or movement and the absolute spatial direction* may serve as similar examples of erroneous representations that were silently accepted. Our ancestors had imagined the Earth as a flat and motionless subject. The moved or motionless objects in relation toward the Earth have been accepted as the absolutely moving or motionless, issuing from such belief. Moreover, all of the objects that were not located on the surface of the Earth and had no support were obliged to fall there by an everywhere-parallel and absolute direction—"downward," to the "natural place" for all subjects. The primitive cognitive problems were raised from such intuitive representation, well-known to us from school course of history: Where is the border of the ground? How much of its edge is far from us? How are the stars fixed in the sky? etc. Long and, in many respects, a dramatic history of refusal from the described absolute concepts (movement and directions) is well known and do not require additional comments. Nevertheless, one lawful question naturally follows up the mentioned instructive historical lesson, not having yet its due illumination:

How we may be confident that in our present representations aren't intuitively established false beliefs, hindering us to estimate the picture of reality without prejudices and rightly?

Close discussion of this question conducts us to the conclusion that the now-accepted belief in our incomparable superiority concerning our ancestors on the mentioned management is not so reasonable.

The many ineffectual attempts to explain the essence of gravitation during long years indirectly point out about presence in our representations somewhat of false belief, having a huge value and directly hindering the comprehension the causal picture of the observable phenomena.

This suspicion and mentioned historical experience oblige us to detail audit of all our initial concepts, somehow connected to gravitation, and to be ready psychologically to their change, even if something will seem obvious and indisputable for us. It is possible to come to a rather-unexpected conclusion for us by accepting such instruction and analyzing the raised question. The whole family of intuitive concepts, originated in consequence of subjective perception of the world, has been present in our representations silently accepted by us from historical times, without conclusive proofs in correctness of them. The forwarded explanation of the nature of gravitation in many respects concerns toward comprehension of values and consequences of our erroneous initial beliefs, which demands patient discussion of their consecution, by starting it from the roots of their formation. We know that knowledge of the material world and discovery of the laws of nature are built on the experimental facts, the quantitative parities of which become out through *measuring*. To discuss the question interesting for us, it will be enough for now to examine a mechanical movement and the circumstances connected to its measuring. For correct and unequivocal description of a movement of an object, it is necessary to define first what relation the movement is considered. What's told is formulated in physics as the "*problem of choice of readout system*" (or base of supervision). It was absolute and unique for our ancestors, as it has been mentioned above, with all difficulties of proceeding from that and, in pair, with some sad consequences. The further history of comprehension of *relativity of all possible systems of readout* and consequential revision of the accepted representations are well known from school textbooks and do not require additional explanations. While registering a movement, except the establishment of system of readout, the observer should use also some minimally required tools and means for locating any of the investigated objects. The certain units of length and of time are accepted to that purpose, well known to us from initial courses of physics. We know that, for example, as a standard of length the platinum core is serving, stored in definitely constant conditions. Clocks or chronometers of different designs are using for measuring the time; the main possibility of them is to generate signals in constant frequency.

We come to the following lawful demand by closely examining the question of establishment of the units of measurement and taking into consideration the former lesson, connected to the system of readout—to a *necessity to revise our primary imaginations in absolute constancy of the used units of measurement.*

Such demand in a first view seems deprived of a sense, and the answer to that seems so obvious, that nobody thinks about necessity of its discussion. However, it is not difficult to guess that our belief in absolute invariance of units of measuring is built only on intuitive perception of the world surrounding us, which reminds us the history with the absolute system of readout. But we should base not on the intuitive impressions only. We have the right to doubt in absolute constancy our units of measurement, without exception the possibility of changeability of their values, issuing from historical experience and the necessity of objectiveness. The complexity here is in the following: With the assumption of changeability of our material standards of measures, we should accept continuous and symmetric, coordinated change of all the material objects; otherwise, the process of change of the surrounded world would be observable for us. The all-round study of the idea of a continuously changing world shows the possibility of its existence that has noticed *Puankare* at his time, for a little bit different occasion. The main significant objection against this assumption consists in other problems. That is the incredibility of preservation in constancy of all primary dimensional parities for the diverse material objects, without single exception, in continuous process of their change. This circumstance, at first sight, is enough, thorough, and significant for interruption of any discussions in this direction in general. However, the marked difficulty may turn out to be the unexpected and significant clue to understanding the essence and principles of formation of a matter in its primary level of structure, being another major problem in natural science. The question concerns the ignorance, to this day, about the physical essence and principles of structure of the fundamental elementary particles, forming all the diverse material objects in our world. The sense of the told consists the following:

Continuously occurred and directly not observable the change of sizes and course of time in our world are possible only in that case if the base of a matter's formation is a unique kind of physical reality, the condition of which continuously changes.

The detailed discussion shows that in this case it will be impossible to find out or deny the change of the surrounded world by means of direct supervisions. It's easier to assure the correctness of this statement if we take into consideration that our standards of comparisons (units of measurements) and systems of readout (frames) may only consist from the same unique kind of physical reality as all the other material objects. Thus, they will change together in coordination and proportionally. Running forward, we shall tell the reader that the idea of uniqueness of a physical reality serving as a base of material world appears extremely fruitful. Very soon, it leads us to disclosure of the physical essence of elementary particles and to the causal explanation of their diverse properties; the second part of the book will be devoted to it.

By continuing those discussions, it will be possible to conclude that we need one more tool in a measuring connected with the movement except those that were already mentioned. The question is about the necessity to transfer information (or a signal) from moving object to the observer that it has not been accepted to speak of at the beginning. It will be impossible to register the moving object's location during measure without such transfer of a signal that is not difficult to understand. Without speaking about it initially the same, we silently were accepting that moving objects are visible to us instantly and precisely there, where they were actually allocated in the moment of measuring. Such imagination had developed because of huge speed of the light serving as a natural information channel, which connects the observer with objects of measuring.

Thus, initially we had supposed without proofs the opportunity to transfer the information in our measuring with infinite speed.

It is possible to guess that from this assumption in our results of measuring will arise some mistakes by values depending on two factors—from *speed of used information canal and from speed of measured objects.* The described circumstance does not deserve our attention for a long time because in daily life we deal with speed much less in relation to speed of light deserving our natural canal of information's transferring. By talking about the above, we are obligated to take care updating the obtained results for recovering the actual picture of observed phenomena, connected with movement, by considering the described factor of distortion in our results of measuring. Under the term of "actual picture" is meant the view of phenomenon, which would be to appear in case of unlimited speed of our information canal. In the quantitative meaning, it will be necessary to add the certain corresponding members in our formulas describing movement and phenomena connected with that. It is possible to guess also that the mentioned correcting members by their structure will look like the parity: *the speed of object's movement / the light's speed.* It is not difficult to understand, too, that these correcting members will be not essential in case of very small values of speeds of the measured objects in relation to a light's speed. In further text, by simple quantitative reasoning, are shown that the consideration of the described distortion's factor leads to the same amendments in measured physical values, which in accuracy correspond with the so-called *Lorentz's transformations of coordinates,* being the basis of Einstein's STR. Thus, the actual meaning of STR consists in a compensation of distortions in measuring, arisen aftermath from defect of the used measuring tool—by limitation of the light's speed. Thereby, the STR more precisely and adequately describes the observed phenomena than the classical laws of movement that are a little bit idealized because of the opportunity of instant measurement (or instantaneousness interaction) silently accepted in that. Thus, Einstein's well-known theory of STR, as per the offered explanation, gets its accessible, clear, logical substantiation being freely from mystical "space-time," which has "special properties" to change the sizes of material objects and course of time depending on speed of movement. This explanation will help the reader to understand, how the numerous mysterious conclusions and logical contradictions in STR come out, over which the physicists and visionaries have argued and speculated for a many years. However, for the sake of justice, it is necessary to emphasize especially that Einstein regards the described deviations of measured values from Newton's laws of movement differently. He calls them "distortions of the actual values," in contradiction to adherents of STR, equating the mentioned distortions to really occurring changes. The necessity to estimate the true cognitive value of "space-time" is very important to us because further Einstein again refers to the same devil—"space-time" to solve the problem of gravitation. This time he gives a new duty to him: "curvature" and creation the gravitational phenomena. On the other hand, the gravitation, Einstein simultaneously proclaims it as something "locally equivalent to inertia" that finally confuses the logical picture in GTR.

In the end of this chapter, we shall give the main conclusions resulting from it, in a brief form, very important for our further reasoning. We shall state them in following points:

1. *The material world, surrounding us, consists of a variety of physical objects, being carriers of energy, capable to independent existence and to interaction, thus, able to be displayed. We have no right to invent new essences in our attempts to understand the laws of nature and are obliged to proceed only from those experimentally established. The observance of such ethical standard is necessary. Otherwise, all our theories would turn into free compositions depending on conscience and imagination of their founders and deprived from scientific objectivity. This requirement frequently isn't respected in modern physics. Logical disorders and insolvable difficulties find place in some fundamental theories resulting from that (examples of said: the mentioned Einstein's theories, quantum mechanics, theory of quarks, etc.).*

2. *Being guided by a principle of objectivity, we are obliged always to take into consideration that all of the possible real systems of readout to descriptions of movement may be defined through the concrete material objects only. Thus, all of the real systems of readout are relative, and there is no opportunity to define the absolute movements and absolute directions.*

3. *The inevitable mistakes occur in our measuring, connected to movement, caused by limitation of speed of used information canal (or by limitation of interaction's speed). Thus, some amendments are necessary to make in our equations describing the movement to correct the marked deforming*

factor. The specified amendments are applied in Einstein's STR—thanks to that, it gives more adequate description of phenomena, connected to movement, than Newton's laws of movement, in which the speed of measuring is considered as unlimited. However, this advantage of STR is obscured by attributing the described divergences from classical laws of movement, toward composed and, from logical point of view, not clear concept—to the "space-time."

4. *We can use only the definitely concrete material objects as units of our measurements, which possess all the necessary properties for that. We consider our standards of length and time as constants by virtue of our intuitive perceptions. However, close discussion of this question shows that this belief can be deeply erroneous. If the sizes of all material objects and frequencies of regular, repeating events in the world would continuously change, by preserving their primary proportions, then such dynamical condition of our world could remain imperceptible for us.*

Thus, the existence and realness for the following beliefs primarily have been accepted by virtue of subjective perceptions:

a) *Absolute motionless system of readout*
b) *Absolute spatial direction*
c) *Absolutely infinite speed of supervision*
d) *Absolutely constant sizes and time*

We have already realized the inaccuracy of the first three concepts from the above list and have managed to get rid of them in the course of development of natural sciences. However, we silently continue to trust in the existence of constant sizes and a rate of time though these beliefs are constructed on the intuition only. Are our meters and clocks actually constant as they seem to us, or they are continuously changeable together with all of the surrounding? We need to solve yet this question, the answer to which is directly related to the nature of gravitation.

THE ESSENCE OF GRAVITATION

The Lord probably has decided what is better to us to see and understand and what is not . . .

Above, we have agreed to be guided by certain rules and principles, the competency and usefulness of which are proved by elementary judgments. By following the mentioned logical instructions, we can come to comprehend the causal mechanism of the group of phenomena named gravitational, which are in our interest. In our routine life, we constantly feel the weight of our own body and of various subjects, which we frequently need to lift or move, except for rare cases, for example, when we fall or we jump from some height. We wish to understand now what force presses us to the chair when we sit, or why the objects fall downward, as soon as they become free from the support.

Such question, probably, will seem a little bit strange or unexpected for many people as the gravitation and falling of subjects are rather ordinary phenomena for us and have been initially accepted as natural things without any special reasoning. On the other hand, in initial courses of physics, it is accepted to bypass smoothly this difficult place, as result of which an impression about complete clearness of all questions connected to gravitation is created today. We have learned to say, for example, that the Earth draws material objects through gravitational field. Afterward, they move toward the Earth by acceleration or press on its surface by the forces of weight. However, the matter is not so clear as it seems by this description. The first difficult question, raised subsequently from the field interpretation of gravitation, is its completely inexplicable property—to force all different kinds of objects without any exception, despite their materials, forms, and sizes to fall to the Earth (or to other source of a gravitational field) *by identical acceleration.* This surprising fact was opened by *Galileo* for the first time and later was confirmed repeatedly with much more exact measuring. For completeness of image of the problem, we shall tell the reader that even the light's beam is falling under the influence of a gravitational field as the various trial subjects. For comparison, we shall remind that the known electromagnetic field, the realness of which is beyond doubts, cooperates with matter by a completely different way—deeply depending on materials of trial bodies. For example, it is possible to be protected from the influence of electromagnetic field by using the screening materials, corresponding to this purpose. However, it is impossible to be getting rid of gravitational field by use of any screens. It is very difficult to explain the described and other features of gravitational phenomena from its field point of view. Except of what is told, it is already possible now to speak about other experimentally revealed facts, testifying against hypothetical gravitational field. The told is concerning a vain attempt of researchers in long years to detect the so-called "*gravitational waves*," the essence of which consists the following:

In accordance to the field representation of gravitation, it should exist the dynamical (or indignant) condition of that also; that is a variable gravitational field. For example, the rotating duple space body should radiate additional variable component of gravitational field, or gravitational wave. The experimenters are trying to register that in long years, applying special "*gravitational antennas*" and detectors for this purpose. However, the mentioned experiments haven't led to any positive result yet, despite the continuous improvement of used technologies. Theorists are inclined to see the reason of negative results in insufficient sensitivity of detectors for gravitational indignations available for today and persistently are planning huge new experiments, or constructing new theories explaining why "the gravitational wave" is so difficult to registering. On the other hand, by careful study of all known gravitational phenomena, it is possible to make sure that among them there are no such effects, which unequivocally and directly would prove the field nature of gravitation. From this point of view, the detection of "gravitational wave" would become unique, indisputable proof of realness of gravitational field.

Nevertheless, as a rule, there are no single proofs in science. Thus, mentioned circumstance causes deep doubts both, over the existence of "gravitational wave" and over the concept of field nature of gravitational phenomena in general. That's why the negative results of experiments to detect the "gravitational wave" are possible to explain not only by imperfection of used devices, but also by inaccuracy of its concept. By rejection of the "gravitational wave," it will be logical to deny the gravitational field also because any of undisputable proofs to its existence as the independent kind of physical reality, simply, doesn't exist. Some opponents against such conclusion can exclaim—how it is right to speak about unreality of gravitational field whereas we constantly feel ourselves its influence, and we see other diverse consequences of that! However, similar objections are completely unauthorized, and it is very important to understand it from the beginning; otherwise, the further development toward the solution of the puzzle will be quite complicated. The unreasonableness of such probable objection is very simple to explain. The matter is that—*the initial facts show here as a proof toward the hypothesis called to explain the same facts!*

The antique Greeks could use such proofs, for example, in benefit to realness of a god *Zeus,* by pointing for that on the fact of thunder and lightning as evidence to a bad mood of heavenly ruler and, thus, to his reality. Similarly, it is possible to show many known phenomena as indisputably proving the realness of gravitational field, such as the tension in a cord under weight of a hanging body, or the sea inflow regularly repeating under influence of the moon, etc. The explanations of mentioned and other gravitational phenomena on a first sight are impossible without the assumption of interaction between material bodies on a distance. It causes the postulation of gravitational field and refusal of other possible explanations. However, we should realize definitely that *the concept of a "gravitational field" is an assumption only, and it isn't yet an experimentally proved physical reality.*

Below, we shall discuss the opportunity of causal explanations of group of gravitational phenomena, without taking the advantage of far-acting interactions through the gravitational field, accepted as the indisputable reality nowadays. We already know from the resulted material that the gravitational phenomena are interpreted in GTR a little differently—as aftermath of "curvature of space-time." It has been also shown that this explanation gives nothing in cognitive direction, but rather on the contrary. The "curved space-time" aggravates the situation even more because of its logical unclearness and impossibility to prove its existence experimentally, unlike "gravitational field," which looks more comprehensive at this point. The "curved space-time" in Einstein's theories only leads away the attention from their rational part, which can help to solve the problem of gravitation as told already in the second chapter.

Now, it's necessary for us closely and patiently to follow to a several mental experiments and reasoning by help of that it's possible to come near to disclosure of the mystery of gravitation. In the beginning, we will imagine one researcher who was born and educated in constantly flying spacecraft. Our researcher in his freely planning ship will have no idea about concepts of "downward," "upward," about gravitation, and other earthly sensations familiar to us. The weightlessness will be normal and natural condition for him, hence, the *equivalence of all possible spatial directions.*

He makes some experiments with the light and finds out that on all directions it passes rectilinearly, without changing its initial frequency (we shall remind that light has electromagnetic, wavy nature). However, the described picture and results of measurements considerably change when the jet engine of the ship switches on. The various unfixed things, for example, that have freely soared in the ship before now start moving promptly toward the bottom of the ship, by *identical, accelerated movement.* After collision with the bottom of a ship, these bodies remain pressed to bottom by certain forces. Repeating all former experiments with the light, the researcher finds out that now it shows different deviations depending on the direction of spreading, which were not present earlier. The initial frequency of light decreases, for example, when the beam is directed from bottom to the nose of the ship, and it increases by the same value on the opposite direction. Some deviation from rectilinear way of light takes place while spreading it on perpendicular direction, in relation to the axis of the ship. As a result, the hit point of light on the lateral wall is a little displaced aside from the bottom of the ship, in comparison to its former place in the case of freely planning the ship, and the way of light seems a little bit curved. There isn't anything unclear in the described results, both for us and for our researcher too. He is familiar with Newton's laws of movement and realizes that all described phenomena occur subsequent to accelerated movement of the ship. This means that the freely soaring objects inside the ship actually continue to keep their initial condition (of a rest or rectilinear movement) in relation to each other, but the ship changes its initial speed with acceleration. Subsequently, the bottom comes closer to the trial bodies with acceleration, and the nasal part of the ship departs from them. The bottom of the ship involves trial objects in accelerating movement too, after their collision. The forces of inertia and counteraction arise between the bottom and trial bodies subsequent to acceleration of the ship, which are equal by their values and opposite by directions. The observable phenomenon with the light also gets its explanation by the same clear reason; during the short time while light passes the distance from the bottom to the nose, the speed of the ship becomes fast a little bit on the same direction. Because of which, the light reaches the nose with little bit reduced frequency, in accordance with the *effect of Doppler* well-known in physics. On the opposite direction of light, its frequency increases by the same value for the same reason. During the passage of light from one lateral wall to another, the ship moves with acceleration some small way. Owing to that, the light's hit point on a lateral wall becomes a little bit moved from its former place toward direction of a bottom. The light's trajectory looks a little bit curved by the same reason. Our researcher makes all necessary calculations and becomes convinced in the exact quantitative consent between results of described three different sorts of experiments:

a) in measuring of acceleration, b) in measuring of frequency of light, c) in measuring of curvature of the way of light.

He also measures forces of reactions between the bottom of the ship and trial objects, and then by dividing the measured values of forces on the earlier measured value of acceleration (on the acceleration of the ship), he defines some individual constants for each object. Those constants characterize *inertness* of trial objects, showing how they resist to external accelerating influence. The parameter of inertness of the object, defined by the described experiments, is accepted to call its *inert mass*.

Let's admit now, one enough extraordinary a case occurs in a life of our researcher. We shall assume that while he was fast asleep, his ship has landed smoothly on a planet. After waking up, our researcher felt the same, familiar force of influence, what he has felt earlier at the accelerating flight of the ship. He again carried out all the above-described experiments and become convinced that all of the trial bodies being free from their support move toward the bottom of the ship with *identical acceleration*, same as before in accelerated flight of the ship. By repetitions earlier describing all of experiments with the light, he found out *all former results and the same quantitative consent between described, different by their sort, three kinds of measuring.*

After these experiments, our researcher quite confidently came to conclusion about the condition of accelerated flight of his ship, based on the own knowledge and repeatedly tested experience. However, by indications of onboard devices, he found out later and become convinced in the switched-off condition of the engine, which made him deeply confused. After coming out of the ship, the researcher seemed to understand what the matter is. The first idea that came to his mind was that the planet on which his ship landed has been moving by corresponding direction with acceleration, as a consequence of all the described events arisen, familiar to him as *inertial phenomena.*

Soon the researcher faced many new difficulties. Going a bit away from the ship and by making all the necessary measuring, he found out that forces of interactions between surface of the planet and trial bodies everywhere have strictly perpendicular direction to the surface of the planet. Because of spherical form of the planet, the researcher thought nothing else except of conclusion that this planet expands with acceleration! Because of such expansion, its surface with acceleration pushed all objects, allocated on it, by radial directions. In this way, it becomes possible to explain all the above-described results, well investigated by him, as inertial phenomena, caused now by expansion of the planet. After such conclusion, our researcher decided to make sure in the expansion of the planet by means of direct measurements. However, all his measurements to define the distance between two points on the planet always show the same ratio that puts him in new difficulties. In the described situation, being guided by the principle of objectivity and having no desire to compose new explanations for a whole group of phenomena well investigated by him, our researcher may only continue the chain of such improbable, but quite logical, conclusions. To explain the absence of any opportunity for direct detection of changes on the expanding planet, he found the following unique conclusion only:

The continuous expansion is peculiarity to all of material objects and occurs by preservation of the primary dimensional parities of all physical objects without any exception.

In the mentioned case, the process of expansion of a material world couldn't be perceivable, except for some attributes, representing from itself the consequences of the accelerated movement—the inertial phenomena causally explainable for us.

Thus, leaving aside our intuitively established bias in the existence of unchanging sizes and measuring units in our world becomes possible the causal explanation of the phenomena referred as gravitation. Of course, at first sight, the logical conclusions of our imaginary researcher seem unbelievable and do not inspire trust. Besides, together with such explanation, many various new questions and contradictions become up, at once seeming completely ineradicable. However, the conclusions of our researcher are fully lawful and consecutive, quite enough for that to be simply not ignored. Now, for a while leaving aside arising various intriguing questions, we can closely trace the reasoning of our imaginary researcher and find out the following small defect in them. The trial objects on the planet have fallen by the radial directions as it has been told earlier. Thus, their trajectories of movement are directed to the center of the planet and are not parallel, whereas in the accelerating ship the trial bodies has moved by the parallel trajectories by a researcher's previous assumption. Thus, this difference ostensibly gives possibility to assert that the phenomena observable in the ship and on the planet are yet not identical and, thus, are distinguishable.

It should be clear that between results of above experiments, made on the planet and in the ship, in the definitely limited volumes of space, it would be impossible to find out any differences, because the sensitivities of used devices are limited and would not be enough for that. Thus, by preservation of described condition, it is possible to accept the equivalence of phenomena in the ship and on the planet. Presented above, the allegation expresses the essence of the *principle of the locally equivalence of inertia and gravitation,* accepted in GTR. Hence, the marked main principle from the point of view of GTR is possible to interpret approximately by the following words:

Two types of phenomena are existing, completely different by their nature—gravitational and inertial, which are possible to consider as identical into limited volume of space (locally)!

Without speaking now about all of the mysteriousness of such statement, we shall only discuss below how this principle is expressed in GTR in quantitative meaning. It is easy to get convinced that the verbal restriction of "locally" in the accepted equality of inertia and gravitation is a result of intuitive perception only, but it is not an experimentally established fact.

The mentioned principle in GTR actually is expressed by accepted *equality of gravitational and inert masses* only, without any of the other quantitative restrictions.

Our reader already knows from previous pages what is inert mass. To understand the sense of *gravitational mass,* it is necessary to look at the above-described experiments on the planet (measuring of acceleration of trial bodies, change of frequency of light and curvature of its way) from the viewpoint of inhabitants of the planet. Being completely confident in immovability and in unchangeableness of the objects that surrounded them, they name the described phenomena as *gravitational,* explaining it afterward of hypothetical "gravitational field" (or by the properties of hypothetical "space-time").

Proceeding from such representation, they measure the interaction forces between surface of the planet and the trial bodies. After, by dividing the measured values of forces on the measured value of gravitational acceleration, they define the corresponding individual characteristics for each trial body, showing its ability to interact with the "gravitational field" (or with the "space-time"). Such characteristic of a trial body, defined by described experiment, they name as its *gravitational mass.* Thus, the above-described numerous experimental results, by inexplicable reason for the inhabitants of the planet (who believe in the existence of unchangeable sizes) show completely exact equality of the marked two physical characteristics— the inert and gravitational masses, describing ostensibly the "totally different properties of matter" according to their previously accepted beliefs.

The equality of gravitational and inert masses is accepted in GTR, naming it as "principle of local equivalence of inertia and gravitation."

Thus, the restriction of "locally" actually didn't have any quantitative role in GTR, remaining as the verbal declaration only (see chapter 2). The identification of gravitational and inert masses is corresponding to the actual identification of gravitational acceleration with inertial acceleration. This means the identification of phenomena named "gravitational" with the phenomena, caused by the accelerated movement. It will be right to suggest the opponents, who feel difficult to trust the above-mentioned statement, to find what are the changes in quantitative parities of GTR, by replacing the principle of "local equivalence of inertia and gravitation" with the principle of their identity. It is possible to get convinced that the change of these word combinations cannot lead to any of the quantitative consequences in parities of GTR. What's told is meaning only that the group investigated in GTR phenomena, carrying the name "gravitational," actually is indistinguishable from the inertial phenomena. Consequently, it is possible to assert: *The quantitative results and productivity of GTR, actually, are conditioned by the dynamical, continuous expansion of the material world.* In the further text, some examples are brought, which show the opportunity to deducing the same quantitative consequences of GTR, experimentally checked up, on the basis of suggested interpretation of gravitation, by means of clear reasoning and by elementary quantitative notions only. The marked solutions evidently show the fidelity of offered explanation of gravitation and clear up the physical sense of GTR. Coming back again to the restriction of "locally," it is possible to get convinced that with the acceptance of general property of a matter, continuous expansion, the problem of "locally" leaves out from the agenda itself. The above-told matter is easy to ascertain by tracing further the reasoning of our imaginary researcher. Accepting in the attention the dynamically expanding condition of a matter, the researcher is obliged in his assumptions to consider the expansion of his ship as well as his measuring units. Subsequently, the researcher concludes that the trajectories of trial bodies falling by acceleration in the ship will not look parallel as he has believed earlier, but its will be little narrowing, same as in their falling on the planet. This explanation shows identity of phenomena in the accelerating ship and phenomena on the planet, thus, the unjustified imposing the condition of "localness" between identical concepts (gravitation and inertia).

By virtue of intuitive belief, the restriction of "locally" in the equivalence of gravitation and inertia now seems to theorists as indisputable and even as not requiring any discussion or experimental acknowledgement. However, at the same time the "principle of local equivalence of inertia and gravitation" seems a bit strange from logical point of view and mysterious to others (for example, for academician *Logunov*). By considering the above told, the necessity of direct experimental acknowledgement of a restriction "locally" in equivalence of gravitation and inertia or denial of it is apparent. The idea of above-described spacecraft may be useful for such experiments as example.

It is possible to assert without quantitative calculations (to have fewer possible reasons to dispute) that

The angular difference between trajectories of falling bodies on the Earth and in the ship will be identical, if acceleration of the ship will be equal to acceleration of free falling on the Earth. The realization of such experiment seems much easier from technical point of view, if trial bodies in the accelerating ship will replaced by two pendulums. The modern tech achievements allow hoping in realization of such experiments to this purpose.

We have discussed only one of the possible ways, conducting to the conclusion about the property of matter of continuous expansion. The "gravitational phenomena," which was not clear for us until now, get an explanation based on this conclusion, not simply for our intuition but quite clear from causal point of view. If such conclusion is true and corresponds to reality, then it will be natural to suppose the existence and to look for new facts additionally testifying to reliability of the revealed explanation. It turns out that corresponding facts and instructions are available, if we'll look on them from the necessary point of view only.

Most of readers probably know about the so-called *Expanding Universe* and about origin of the substance afterward of a "*big explosion.*" The brief description of circumstances to its establishment and its concept consist in the following. In the beginning of past century by supervisions in spectra of light, radiating from remote space sources, by *Hubble* has been opened a displacement of spectral lines from their normal position on the side of red color. This displacement of spectral lines corresponds to the change of frequency of light, in this case—to its reduction *(red shift)*. Moreover, it has been established that the described displacement of spectral lines and the reduction of the frequency of light are directly proportional to the distance of observable object *(Hubble's law)*. The following two different interpretations explaining such change of the frequency of light are under consideration in physics until now (reader can already understand the circumstances for such duo explanations). The first of them explains it as aftermath of causally well explainable in physics the *Doppler's phenomena,* caused by a relative movement between the source and the receiver of light. The second one connects the change of frequency of the light with the influence of "gravitational field" on it, the causal mechanism of which for theorists remains obscure until now. Theorists have come to conclusion about the dynamically expanding condition of a Universe by issuing from described fact of red displacement a light and preferring the first explanation by some reasons. Except of the Hubble's change of frequency of light, there is also another experimentally established fact (the so-called *background radiation*) and some theoretical conclusions to the benefit of expanding condition of the Universe. The speech concerns the works of *De Sitter* showing that the Universe can exist in expanded condition only, and to the *Freedman's* works concluded by analysis of GTR and showing that Einstein's initial representation of stationary condition of the Universe contradicts his theory. This contradiction was eliminated with an assumption of expanding the condition of the universe. The fairness of mentioned conclusions should be quite clear for readers under the above explanations of identity, the concepts of inertia with gravitation, and from interpretation of physical sense of GTR as actually based on the property of *continuous expansion* of matter. With acceptance of expansion, the conclusion about the origin of our material world from an insignificant small volume of space at a definite moment of time is to follow, as the afterward of an unimaginable cataclysm from the completely unclear for us condition of a substance. The described mental picture has adopted the name of *Big Bang* among theorists. Above, we have presented the existing preconditions leading us to the concept of expanding Universe in general. Nevertheless, it is also necessary to speak about one unpleasant circumstance generated with this representation. According to the nowadays-accepted view, it is accepted by theorists that the remote space objects—the galactic systems and galaxies in them—separately have been removing each from others according to Hubble's law. The galaxies themselves, on the example of ours, are considered as invariable in their sizes, except some developing ones. The difficulty here is that no evidence of a difference in structure of a different scale's groups forming the different types of space objects in the Universe. Same as the galactic systems, the galaxies and the stars with their planetary systems also exist or develop based on the general principles of their structure and behavior. Thus, there is no answer to a natural question: why the galactic systems and galaxies should be moving away from each other, but distances between stars and between planets should remain constant? It is necessary to mention that the raised question doesn't deserve attention in modern theories and in numerous scientific discussions concerning space and gravitation. It is easy to explain such reality because of the above-described formalistic approach in modern research works. Like in many other questions, in this case also, the theorists put a task of establishing the existing facts and ways for their quantitative description without paying attention to such "trifles" as logical or cognitive disorders by leaving it for the future in the best case. Meanwhile, it's one serious cognitive problem demanding its reasonable answer from the theory of expanding Universe. It is possible to find out the one methodological inconsistency in the accepted modern concept of expanding universe by detailed discussion of the above-described situation. There are questions in that the theorists have judgment about the expansion of galactic systems and Universe based on red shifts in the spectra of light, while the existing belief in constancy of sizes of the galaxies, stars, or planetary systems are constructed by results of direct supervisions. Thus, in this case, two principal different ways are applied to establishing the facts and to getting their estimations. We can't exclude that the answer of the raised question is joined with nonequivalence of these two methods. If, for consistence, we accept the red shift as a general attribute of speed between source and receiver of light then all of the above-mentioned space objects become equal in their rights and indistinguishable from each other in this respect. The red shifts in spectra of light generally always take place at its passage through the environment, depending on the distance and from density of substance; it is well-known to experts in this sphere. According to nowadays-accepted approach, the red displacement of the spectra of light is considered as attribute of speed in one case, in scales of Universe, but in the other cases, in scales of galaxies, planetary systems, etc. The same phenomenon is attributed to "influence of gravitational field on the light." It is necessary to note, unfortunately, we can't put a requirement to judge about the expansion of all the mentioned space objects in the general way—by direct supervisions because of supervisions of incredible distances and intervals of time. Thus, the picture of expanding Universe nowadays seems rather subjective: *The Universe expands in scales inaccessible for our direct supervisions but in observable areas that the sizes of objects and distances between them remain constant!*

The described picture does not maintain any criticism and does not require many comments from the viewpoint of logical or scientific objectivity. To find a way out from such doubtful situation, it is only possible to choose one of the two opposite directions—either to refuse the concept of expanding Universe at all, or to consider the expansion as general property of all materials against our intuitive perceptions. The expansion of the Universe has been accepted after long discussions, under the pressure of some circumstances, with some of which we have already gotten familiar above. Its refusal would mean denial of the mentioned arguments, which would be nonscientific and unauthorized simply. Hence, we have the second way only, sacrificing the intuitively accepted belief in the existence of the absolute constant sizes in our world and acceptance of expansion as a general peculiarity of all of the material.

It is also possible to find other simple reasons in the benefit of extending condition of a matter, proceeding from some existing circumstances. For example, by careful study of all the initial principles to construct GTR, it is possible to get convinced in absence of whatsoever restriction among them carried size-scale character. Hence, the above-mentioned *De Sitter's and Freedman's conclusions, acquired from equations of GTR, are right in any scales with equal competency, in relation to Universe, as well as for the galaxies, planetary systems, and to a directly observable material objects also.* This is another and almost independent reasoning, simply bringing to the concept of the expanding world.

In continuation to this question, it is necessary to mention about one important circumstance also, testifying the correctness of the offered interpretation of gravitation. Speaking about Hubble's expansion and the majority of votes accepting its correctness, at the same time, the theorists do not wish to pay attention to circumstance that both *the quantitative descriptions of expanding Universe and gravitational phenomena proceed from the same equations of GTR.* The gravitational phenomena and expansion are accepted to look now as directly not connected to each other two independently and contrary factors, for example, in modern attempts to defining the future of Universe. By the accepted imagination now supposes that it will be possible to define which of two "contradictory" factors—the speed of expansion or the gravitation—will be prevailing in the further behavior of the Universe, by correction of values of Hubble's constant and of average density of substance. The expansion either will last eternally, or after a certain time, it will stop; or it will be replaced with the compression of substance, etc. To show the incorrectness of such concept to this question, we shall allow ourselves the following comparison. Let's imagine a researcher, who is unfamiliar with the Newton's third law and tries to define by exact experiments, which one of the two forces is more, the force of action or counteraction. The raised modern problem to define the future of the Universe by its meaning is rather similar to the described researcher's task. Our reader already knows that gravitational phenomena and attributes of expansion of the Universe are different displays of the same process. The above conclusions of de Sitter and Freedman's served as quantitative proofs that the descriptions of expansion of the Universe and group of gravitational phenomena by the same equations of GTR.

Thus, *the factors of expansion and gravitation should be in accuracy equal to each other by their quantitative parities, as the action and counteraction in the third law of Newton.*

The matter is in that the told actually has proved in accuracy in the framework of today's possible measuring. The theorists cannot yet define which of the mentioned two factors is more by present estimations of average density of substance in the Universe and Hubble's speed of expansion. They still do not think that it will always be impossible despite the continuous improvement of methods of estimations. This conclusion proceeds from a clear reason that the marked factors are interconnected and are quantitatively identical. The given statement can serve as the other experimental direction to prove the suggested explanation of gravitation in the future. Thus:

Above, we have studied some reasoning leading to a conclusion of dynamically expanding condition of the material world. The gravitational phenomena, known to us, get their causal interpretations based on this conclusion, as clear consequences of continuous expansion of a matter. This representation deeply contradicts our intuitive perception of constant sizes and demands the certain efforts to overcoming our biases. However, similar problems are not new in the history of natural science. The former ideas about absolute movement and absolute direction serve as examples for such difficulties, connected to comprehension of inaccuracy of our initial concepts. All difficulties, connected with comprehension of the essence of gravitation, in many respects, are conditioned now with the necessity to refuse the next false concept—the representation of absolute constant sizes and time. Having necessary resoluteness and by overcoming our intuitive sensations, it's possible to understand gravitational phenomenon as comprehensive aftermath of a unique reason—dynamically expanding condition of our material world, unknown to us before.

THE QUANTITATIVE DESCRIPTION OF AN EXPANDING WORLD

Here I will just apologize; we cannot bypass more without of little bit mathematics

The concept of extending material world depicted in the above lines had opened the causal interpretations of gravitational phenomena. However, the offered explanation of gravitation, even persuasively, may remain only as one tempted history if not to bring concrete quantitative conclusions, corresponding with the experimentally established facts. Hence, it is necessary for us to be patient and to follow some quantitative reasons, showing the ability of the introduced concept and the conformity of the presented picture to a reality. In the above text, it was underlined and argued that continuous expansion of the material world by its quantitative parity, actually, completely coordinates with GTR and expresses its causal essence. This means

The mathematical structure of GTR and its equations are expressing quantitative parities of physical values in expanded world.

The following interesting circumstance, shown in structural features of formulas determining values of any gravitational effect in GTR, may serve as an additional proof to correctness of the above. The final expressions for gravitational effects, as it's easy to see, together with numbers and geometrical constants contain only such members, which by their measurements can be presented as degrees of parities of a certain speed to a speed of light. Without having any knowledge of the nature of gravitation, it's possible to make the below conclusion based on the mentioned fact only:

The certain movement causes all of the different gravitational phenomena; thus, those are the aftermath of somewhat of dynamical process continuously occurring.

We have already studied from previous material that this process is the continuous expansion of a matter. Taking into consideration given explanations and mentioned reasons, we could be satisfied by the existence of GTR and other alternative theories, by using their quantitative parities. However, the accessible, clear, causal interpretation of the nature of gravitation gives us an opportunity to do little bit more and in much easier way compared to existing formal mathematical theories, about which it's possible to judge by reading the below chapter.

Proceeding from the aforesaid and considering the popularly accessible narration of this book, we shall be briefly examining the minimal necessary quantitative reasons with the use of elementary mathematics only. However, some information and comments about quantitative parties of the theories of STR and GTR are necessary as the primary familiarization with the studied problem.

Generally, the task of quantitative description of the any phenomena is meaning the formulation of a kind of mathematical parities, taking place between physical values in that phenomena correctly corresponding to an experimentally established fact. According to nowadays-accepted concepts, a high-rating formal mathematical theory should solve this problem in general. In particularly, the good theory should provide such system of equations, which, at least in the basic plan, allows solving any problem in the investigated area at any reasonable primary conditions. The GTR corresponds to the stipulated requirement in description of gravitational phenomena. As per the offered explanation of gravitation—*for quantitative descriptions of gravitational phenomena it is necessary to take into consideration two factors only—the finiteness of speed in measuring and the dynamically expanded condition of a matter.*

As we know already, the specified two factors are actually considered in equations of GTR, thanks to the quantitative conclusions that correspond to the experimentally established results.

Nevertheless, the problem of the practical use of correct equations follows after their formulation. The matter is that the requirement of consideration of the mentioned factors in the general description of movement leads us into serious technical difficulties. To understand mathematical complexity of this problem, we should first recollect how movement is described in classical mechanics. As a general task of the problem, it's demanded to define the location, the values of speed and acceleration of a physical object, with known mass and sizes, in an interesting moment of time, with known laws for the forces influencing that. We know that *three-dimensional description of movement is applied for this purpose in classical mechanics, with the using of the Deckard's, spherical or cylindrical coordinate systems of readout, with an assumption of an independent course of time and an opportunity of instant measuring in that.*

From this brief description, it is already possible to imagine what mathematical complexities can be risen while solving such problems, when the acting forces on the object are many and variable, depending from location (from the coordinates) of moving body and from the time too. The adequate and causally proved system of equations to describe the movement in general has been formulated in classical mechanics quite long ago. However, their practical implementations are always limited with its use in special conditions only. For example, when the forces are one or two, and they change under rather-simple laws, etc.

To understand the essence of mathematical structure of GTR and its complexity, it is necessary for us to study first some of the questions connected with the choice of systems of readout in descriptions of a movement.

The concrete location of the observer is established simultaneously with choice of the systems of readout in classical mechanics. The results of measuring depend on that in many respects. It is obvious that in this case the used equations describing movement will correspond with the results of measuring in the chosen system of readout, having other possible expressions in the other systems of readout. The description of a movement of the planets can serve as an example of the above said. As we already know, the movement of the planets occurs according to *Kepler's laws,* by the elliptic trajectories, one of their focuses almost coincides with the location of the central star. Nevertheless, to confirm the correctness of that by means of direct supervisions, the observer has to be located on the central star, or in other words: his system of readout should be connected with the central star, which in frameworks of the investigated phenomenon is possible to consider as absolutely unmovable. For the real observer, allocated on a planet, the movements of the other planets will seem with rather-confusing trajectories at first sight having nothing common with the Kepler's laws. The *Copernicus's* big merit consists in proving the utility of mental transition in the Solar system. Thanks to that, it was possible to see out a strict rule in confusing trajectories of the planets and to formulate the laws of their movement. Based on the given example, it is possible to judge about the value of correctly chosen systems of readout in descriptions of a movement.

However, the requirement of choice of the good systems of readout is a technical question only as the facilitation to the description of movement hasn't the principal value.

The sense of this statement can be comprehensible from the continuation of the above-considered mental experiment. Let's assume now the observer shall decide the next practical question: How to describe the movement of the neighboring planet from the own system of readout, connected with the own planet? To define the location of the other planet in relation to his own, at the required moment of time, the observer can do the following. In the beginning, by the Kepler's laws, he can specify the location of the own planet as well as the planet interesting for him, in relation to the central star. Then, without special complexities, he can define the distance between two planets and there, the mutual position between them also. Thus, to simplify the problem, the observer in the beginning chooses *intermediate mental system of readout*, connected to the central star and convenient for calculations, and then, for getting the final answer, he makes transition from mental to a real system of readout, connected to his planet. From the described way of solving the problem, the basic opportunity to solving it in the generalized form also is obvious. For this purpose, it will be necessary to use only the generalized mathematical designations and, by leaving the intermediate calculations, to come to the final expressions. The system of equations obtained in such way would express all the above-described operations on transition from the one system of readout to another and, based on the Kepler's laws, would connect the parameters of two planets and the central star. The equations, made in such way, would already have universal character, allowing to apply that to solutions of similar problems for any pair of planets, without thinking of systems of readout or about their choice. For the reader, having mathematical knowledge and skills of its use, it will be clear that by their structure and by quantity of an independent argument, the acquired equations will be much more complex in comparison to the previous ones, which were using the convenient systems of readout, hence, dividing the general problem into separately solvable parts. However, by virtue of universality, the second way of describing the phenomenon is generalized and preferable in modern formal theories. The equations of generalized description of physical phenomena, without using specially chosen systems of measuring, are accepted to name *covariant equations*. We already know from previous text that the equations of GTR, actually, are the equations of a movement. They are different from the classical equations first by their *covariant* character, which causes the complexity of their structure in many respects. Moreover, in the equations of GTR, the finiteness of speed in measuring is accepted, connected with the movement (by accepting the "*Lorentz's structure of space-time*"). Actually, the general expansion of a matter and the change of the sizes during measuring are also accepted in GTR, as explained in the text above (that consists accepting the "principle of equivalence of gravitation and inertia," or the "curvature of structure of space-time").

From above description, it is possible to judge the essence and technical complexity of the practical use of the equations of GTR, in case of a generalized problem. Hence, our reader should not be surprised that the basic solutions of the equations of GTR are not found yet and not many people hope to get that. All of those which are possible to do by using the equations of GTR till today are the description of the effects for the particular conditions only, allowing approximate methods of solutions (see *Shwarthshild's particular solution*).

For complete estimation of the equations of GTR, it is necessary to add to the above said information the one remarkable feature of them, having the very important cognitive value. Question in that—for an opportunity of existence of the material world in a harmonious, coordinated, expanding condition occurring imperceptibly for us, the substance should possess all of the necessary properties for that, some of which could be unfamiliar and unclear for us before because of our primary false beliefs. However, these properties of matter should necessarily be exposed in a theory adequately describing the expanding world. Some laws of nature and properties of matter could seem to us before as the independently existing, separate facts while representing the world in static condition. Nevertheless, for such opportunity of self-coordinated, dynamically expanding condition of a material world, a harmonic intercommunication between many of the various properties of matter is required. We can be surprised only by Einstein's genius, actually, ignorance about causal essence of gravitation but realizing by intuition (or from some abstract reasons only) the importance of a harmonious communication between various properties of matter and the necessity of its consideration in his theory.

One of the examples of the revealed new features of matter is Einstein's well-known parity of mass and energy contained in it:

$$E = m\,C^2 \qquad\qquad (1)$$

Where: *E:* the energy, *m:* the mass, *C:* the speed of light.

This law extremely is important in the modern physics. However, it remains not quite clear in cognitive sense until now. Why does it take place? What kind of relation may exist between the mass of a matter and speed of light? It is possible to set a lot of similar questions, but precise answers to them do not exist. That may not be surprising, as by modern methodology is not accepted to pay much attention on the logical questions or on the reasoning.

With the accepted methodology, the physicists have a task to establish quantitatively authentic descriptions of facts and continue counting on further. However, in the revealed parity, except of its quantitative site, it also contains the cognitive meaning in huge value. In particular, this formula, by virtue of its universality, is possible to consider as a weighty instruction to the following:

All of the various material objects shown in diverse forms and conditions in the root have only one general basis. Moreover, the presence in that of the electrodynamics constant (speed of light) tells us that *the general and unique basis of matter has electromagnetic nature.* Thus, Einstein's well-known formula— (1) can be perceived as quantitative acknowledgement to a logical conclusion about uniqueness of a physical reality, serving as base of matter, which has been mentioned in above pages. In this regard, it seems pertinent to refer to Einstein's known statement about "falseness of the division of a matter from a field, after the assumption of the equivalence of mass and the energy" [L-2]. The other important consequence of GTR, also not deserving due attention among theorists, is *the definition of concept of time as density of energy:* [L-8]

$$f = 1/ t \sim w \tag{2}$$

Where: f: the course of time (or the frequency), t: the interval of time, w: the density of energy (sign $\sim$ means direct proportionality).

We shall see a little later that physical sense of a parity, on a sight rather mysterious, is surprisingly simple, and very soon it can lead us to deducing Newton's well-known law of universal gravitation. Considering parity (2) and the equation (1), it is possible to conclude:

$$f \sim w = m \, C^2 / v \sim \rho \tag{3}$$

Where: v: the volume of substance, ρ: the density of substance

The parity of (3) is showing that the average ratio of event's frequency (or the course of a time) is a property of matter, and it directly is proportional to its density. Considering as a classical and the simplest example of a material object—the ideal gas—we shall try now to design a real clock on its basis. For this purpose, it is possible to count up, for example, some concrete number of the molecules of a gas, passing through a certain section in the volume occupied of a gas, accepting the same number as the unit of time's interval. Supposing that at changes of the condition of a gas, the exchange of its energy with the environment does not occur; from respective section of physics, it is possible to make sure that frequency of passage of the molecules of a gas through the unit section is really directly proportional to its density. This conclusion directly confirms the parity (3). By comparing this parity with the equation of the energetically isolated gas, it is possible to conclude:

$$f \sim \rho \sim P \tag{4}$$

Where: P: pressure of a gas

Thus, the consequence (3) in GTR, actually, represents itself a little bit of the other expression of the well-known law of the condition of gas. With primary representation of matter in globally static condition, the average frequency of regularly repeating events in the material world is also considered as the constant or absolute. Thus, the necessity to relate the concept of time to properties of matter remains not realized and not demanded. Nevertheless, speaking now about expansion of a material world, we are simply obliged to take into consideration that every possible kind of real clocks can have only material basis.

Thus, we should ask a natural question: How does time be changed because of general expansion? Hence, comprehensive reasoning and equations of GTR are showing that

The average frequency of regularly repeating events continuously should be decreasing in the expanding world.

This conclusion from GTR corresponds to known properties of a matter and looks rather clear in the context of the concept of expanding world. The inaccuracy of abstract concept of time, running similarly everywhere and having no direct relation with the matter, is obvious from this definition.

While studying the phenomena related to expansion of matter (gravitational phenomena), the concept of "time" can be considered only as concrete local value, depending on density of the substance, contained in that spatial volume at the rate of which a problem is considered.

Based on what's told, we can find out how to apply the parity (3) while solving specific problems, related to the gravitation. At studying the phenomena occurring in spatial volumes, in scales of which the distribution of substance can be considered uniform, it is obvious that the parity (3) can be used without any modification, accepting the density of substance equal to its average ratio in the observed volume of space. While solving problems related to the separate taken material body, to keep the parity (3) for some point in space being far from the observed massive body, the course of time and the density of matter should be considered:

$$f_R \sim \rho_R = 3\,M\,/\,4\pi R^3 \qquad\qquad (5)$$

Where: ρ_R: the density of substance in observed point, M: the mass of body, R: the distance of observed point from the center of mass.

For the several material bodies, the course of time and the density of substance in the observing point will be defined according to the parity:

$$f_{RE} \sim \rho_{RE} = 3\,(\textstyle\sum M_i)\,/\,4\pi R_E^3 \qquad\qquad (6)$$

Where: ρ_{RE}: the density of a substance in the observing point, $\sum M$: the summary mass contained in a spherical space by radius: R_E.

Thus, the parities (5) and (6) are showing that in the gravitational phenomena, connected to separately taken gravitational bodies, the value of density of distribution of substance, or by definition—the course of time, depends on the distance of the observed point from the center of a mass for all of the bodies. In case of equal spatial distribution of a gravitational substance, the course of time is constant; hence, it does not depend on location. It is easy to get convinced that while increasing the numbers of the gravitational bodies and spatial volumes, the parity (6) automatically is transforming into the parity (3).

It is necessary to bring here some quantitative reasoning also to explain the role of the speed of light in measuring, about which it has mention in the previous pages. The consideration of a finiteness of the speed in measuring, actually, is expressed in the use of *Lorentz's transformations of coordinates* in modern theories of "space-time" and gravitation, in contradiction to the classical physics, in which *Galileo's transformations of coordinates* were used, with an assumption of instantaneousness of the measurement process. It is possible to show the fairness of this statement through the below described mental experiments.

Let's assume that the experimenter is measuring the length *(l)* of the same rigid core, in conditions of its rest and in its movement with a speed *(V)* in comparison to the observer and to the source of the light *(S)* (see fig. 1).

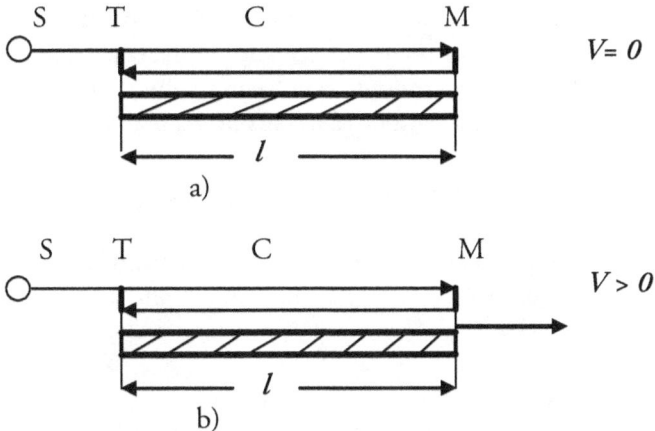

Fig. 1. Distortion of results of measuring in consequence of the movement of the measured object

For this purpose, he establishes the timer *(T)* connected with a photocell on the left end of the core and the mirror *(M)* on the right. The timer starts working from the moment of hit of the light to the photocell from the left side, and it stops at hit of the beam from the right side. In the first experiment (a), the indication of the timer will be defined by parity:

$$t_1 = 2l / C \qquad (7)$$

In the second experiment (**b**), the indication of the timer is possible to find out based on the classical laws to the summing up of the speed. The following value will come out:

$$t_2 = \frac{l}{C - V} + \frac{l}{C + V} = \frac{2l}{C\,(1 - V^2 / C^2)} \qquad (8)$$

Thus, the parity of results of two measuring should be:

$$l / l' = t_1 / t_2 = 1 - V^2 / C^2 \qquad (9)$$

Hence, in accordance to "old" classical laws of physics, some change of length for the moved object should appear in the results of measuring (9). It will not be difficult to guess that in measuring the speed of light the expected ratios at two described experiments will correspond to the same parity (9) also. However, some experiments with the light (*experiments of Michelson and Morley*) was showing the invariance of the light's speed, independent of movement of the system of readout, which was so confusing for the physicists in that time. We already know that the formal concept of the "space-time" was inserted in STR to explain such results of experiments. Such properties as change of the sizes of material bodies and course of time in the moving system of readout have been attributed to a "space-time," in conformity with the following parities (*Lorentz's and Fitzgerald's reductions*):

$$L'/L = f'/f = T/T' = \sqrt{1 - V^2/C^2} \qquad (10)$$

Where: L, T, f and L', T', f' are the units of length, of the interval of time, and of its frequency, accordingly in the motionless and in moving systems of readout (considering $T = 1/f$).

According to this explanation, the difference in the measured results (9) can be attributed partially to reduction of the unit of a length, and partially to the decrease of the frequency of time (or to the increase of the interval of time) in the moving system of readout:

$$L' f' / Lf = \sqrt{1 - V^2 / C^2} \times \sqrt{1 - V^2 / C^2} = 1 - V^2 / C^2 \qquad (11)$$

In such a way, the differences between measuring and expecting results were explained.

It is possible to judge the true values of Lorentz's transformations of coordinates and about the concept of a "space-time" based on the resulted example. However, it's not necessary to move away from classical physics and to composing new things for causal explanation of invariance of the light's speed independent from movement of the readout systems, in the author's opinion.

For this purpose, it is necessary to pay attention first on the issue that in the above-described mental experiment the speed of light and object's speed are considered here as equal categories. However, the speed of light is a speed of wave process or excitation, which is so different from moving material objects. As an example, we can recollect that the electrical signal passes in wires with the speed close to a speed of light, while the speed of the material carriers of the electricity itself has insignificant value only.

Thus, the question is, how right is the summarization of the speed of excitation with the speed of material objects? By author's viewpoint, a wrong assumption has occurred in the judgments at beginning. The matter is that, while comparing the speed, we have looked the objects as the material points, not having own sizes. However, such assumption principally is incorrect in relation to a wave process (we can't assume the wave's length equal to zero, for example).

It's possible to explain this question with the use of *Young's* well-known formula:

$$C = \lambda v \tag{12}$$

Where: λ, v, accordingly, is the length of the wave and the frequency of light.

Assuming the observer's system moves in relation to the source of light, it is possible to define the frequency of light measured by him according to *Doppler's effect*:

$$v' = v\,(1 + V/C) \tag{13}$$

Where: V is the speed of approach of the observer's system to the source of light. The measured length of the wave will be defined according to that effect:

$$\lambda' = \lambda\,C/(C + V) \tag{14}$$

The modern and past generation's physicists also have never seen any of mystery and inexplicableness in Doppler's effect. By accepting the speed of light as a derivative value, determined through λ, v, in the moving system, in agreement with (12) and from Doppler's phenomena, it may be defined:

$$C' = \lambda'v' = v\,(1 + V/C) * \lambda\,C/(C + V) = v\lambda = C \tag{15}$$

The same reasoning is right in case of both movement of the observer from the source and movement of the source in relation to the observer. Thus, this simple reasoning based on a light's wavy nature had shown that the measured value of its speed would be constant and independent from relative movement of source and receiver.

The one interesting acknowledgement to a quantitative equivalence of the Lorentz's transformation of coordinates with the distorting effect in measuring, caused by limitation of light's speed, is possible to find in modern directories and dictionaries of physics. It is clearly written there, "the changes of sizes of moved objects will be invisible because of various delays of light's signals coming from its different points." [For example, the *Kobzarev's* encyclopedic handbook.] (*It actually means that the Lorentz's change of length and the distortions in measurements, caused by finiteness of light's speed, are equal by their values!—author.*)

The fact of constancy of light's speed finds a new sense in the context of the offered concept of an expanding world. The concepts "rest" and "movement" becomes subjective determinants only, not connected directly to properties of matter by issuing from the viewpoint of the dynamically expanding condition of that. It means only those special cases when parities of distances between material objects and their sizes are preserved or changed while its values are continuously varying. Thus, in the expanding world, there does not exist any principal difference between resting and moving material objects relative to each other that expose in the invariance of light's speed, in relation to all sources and receivers of that. There are many written pages, and discussions are continuing to this day about light's speed. We'll be limited by the above-stated reasons, mentioning only that in our further reasoning we will proceed from accepted principle of a constancy of the light's speed. Based on a previous explanation, *we shall consider the value of the light's speed as an inevitable factor in our measuring, not mentioning more about "space-time" in future reasoning.*

Except of the above, it is necessary to have some additional explanations about the used method of description for the sake of clearness of further reasoning. It's needed to tell first that it becomes not obligatory to use the formal mathematical device of GTR and its equations to solve some concrete gravitational problems, the causal essence of which is already known to us from previous pages.

It becomes possible to avoid the generalized, covariant description of phenomena by choosing the convenient systems of readout in each concrete case, which so much facilitates the solution of the problems. Moreover, the three-dimensional spatial description becomes also nonobligatory in some problems, considering the gravitational effect as a function from one or two coordinates only. The described methodological clauses do not change anything in results of calculations (convincing examples of which are abundantly in classical mechanics). It only facilitates solution of problems, bringing them sometimes to the level of elementary mathematics only. In connection to what's said, it is preliminarily necessary to tell also some words about defining the system of readout used by us while describing the gravitational phenomena. According to the initial principle, the entire material world expands with preservation of primary dimensional parities of all the objects. Hence, only an imaginary mental system of readout can be used to describe the expansion of matter, with assumptions of its absolute immovability and the presence of absolute constant measuring units in it. We shall relate our mental system of readout to special points, which can be considered motionless within the framework of studied problems (by the example of the Solar system of readout, at the description of the movement of planets). We will accept the measuring units in the mental system of readout as equal to those in the expanding world, at the initial moment of counting. It's possible to judge the meaning of such approach on the example of *Copernicus's system of readout.*

As we know, the mental system of readout by Copernicus has been connected to the Sun, which may be considered as allocated in the center of summary masses of the planetary system and motionless, in a framework of description of the movement of planets. In the mental system of readout, constructed by us, in addition to the above-mentioned conditions (central symmetric position and immovability), we shall also assume an opportunity of instant measuring and the presence there of the absolute, constant units. Despite the spacious descriptions, the logic of our subsequent actions is easily explainable. In case of a Solar system of readout, we mentally had passed in the best system of readout that helps us to solve the problem of how we will look from there the movements of planets. After taking into consideration the movement of our planet, in consequence of which the trajectories of space bodies seem deformed, we may decide already how the movement actually will seem to us from our planet.

In case of the expanding world, we already know not only about the movement of our system, but also about its continuous expansion and about the absence of an opportunity of instant supervision (measuring). That's why, now we should build the best mental system of readout, taking into consideration more deforming factors already and corresponding amendments to that. Based on described explanations now, we can start concrete calculations and deduce some conclusions. At the beginning, we shall examine an expanding, homogeneous spherical body. The mental system of readout, we shall connect with the center of the body; and the single axis of coordinate, we shall direct by its radius (fig. 2).

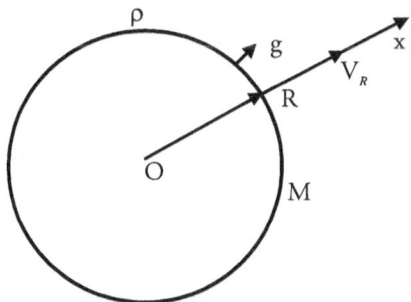

Fig.2. The expansion of a spherical body

We shall designate: M: the mass, ρ: the density, R: the radius of the body, V_R: the speed of departure of the point allocated on the surface of the body, g: the acceleration of that point.

According to preliminary condition about preservation the proportions of all material objects during its expansion, it is possible to write:

$$V_R \sim R \qquad\qquad (16)$$

From (5) it is possible to define an interval of time corresponding to distance (R) from the center of the body:

$$t = 1/f_R \sim 1/\rho_R \sim R^3 / M \qquad\qquad (17)$$

Ensuing from parities (16) and (17), it is possible to define the acceleration of expansion corresponding to this distance:

$$g = dV_R / dt \sim M\, d(R) / d(R^3) = k\, M / R^2 \qquad\qquad (18)$$

Where: k: certain factor of proportionality determined by experiment of *Cavendish*: $k = G$ (it is the gravitational constant). Symbol **d** is the *differential* of values.

This expression is showing that the expansion of matter, in conformity with the stipulated conditions, may be occurring by acceleration only.

The parity (18) can serve as a conceptual substantiation of Newton's law of attraction known as the law of universal gravitation, deduced by him from the analysis of *Kepler's laws*: thus, on the basis of observable results only (but not on the basis of conceptual reasons as in the above.)

The representation of accelerating expansion of a matter generates many intriguing new questions having cognitive character and demanding explanations. However, the similar difficulties are mostly connected to our former intuitively generated representations. The conclusion about continuously growing speed of the expansion is one of the examples of such problems as a logical consequence of the accelerated character of expansion. The offered explanation, though not fully ordinary, consists the following:

The acceleration of expansion is not caused by increase of speed as it was represented in the classical mechanics, but it is conditioned by continuous reduction of the course of time, caused by decrease of density of the matter aftermath its expansion.

It would have been impossible to guess or think about such explanation in framework of concepts of classical mechanics where the course of time and density of the matter were silently considered as the invariant values.

Another intriguing question, rising from replacement of Newton's gravitation by the expansion of matter, is the opportunity of equilibrium existence of the planetary systems with the central star and the possibility to explain the heavenly mechanics, in general.

It is necessary to confess frankly the difficulty of having a convincing interpretation of the observable picture of movement of heavenly objects in descriptive language from position of expanding world, by excluding far action of gravitational forces. However, this question does not differ from others in a quantitative point of view, as all the similar problems will go back to the unique principle—to the above *identity of concepts, the inertia and gravitation.* The author considers it to justify stopping the one characteristic question of similar problems rising from denial of field nature of gravitation.

As it was accepted earlier, each of the planets was "feeling" through the gravitational field where the central body was allocated and was moving around it. How can the planet "know" now about location of the central body and "define" his way, in case of the absence of gravitational field? It is necessary to start from the big bang to understanding the causal essence of such phenomenon from a new point of view. It is easy to conclude that all of the materials forming any of the space system have been concentrated together at some stage of its evolution, according to the concept of expansion. The different objects of this system separated from the general center of masses by somewhat a reason, and now they are continuously departing by spiral trajectories (if looking on that not in an expanding system of readout) in coordination with general expansion of matter. It seems to us in our real world that the bodies, being free from force influences, are in rest or moving rectilinearly evenly (by the first law of Newton). Nevertheless, the relative speeds of free bodies, actually, are directly proportional to distances between them. Thus, to describe the free movement in the not expanding mental system of readout, we are obliged to refuse the first law of Newton ($V = const$) and replace it with the Hubble's law $(V = HR)$. However, such replacement of our fundamental law of movement leads to other vital consequences too. In particular, the free movement of material bodies becomes an accelerated movement in the not expanding system of readout. Thus, it gets new quantitative properties about which we could not guess in our real expanding system of readout.

According to our former representation, the free movement contains the information about current location, speed, and direction of a moving object only. In the not expanded mental system of readout, where we wish to describe the same movement, the freely moving object carries also the information about the initial moment of movement. Since the movement already becomes accelerated, it becomes possible to count from its current parameters—when and where it has begun. Thus, it becomes clear that the freely moving body may "remember" the place of initial point, where its movement started under the influence of an initial impulse obtained by somewhat a reason.

According to this explanation, the expanded body can also "remember" short-term external force influences on it and make complex fluctuations, in relation to the central point of its expansion, etc. To easily create an image of the told, for example, we can consider that the body composes one system from many pendulums that are connected all together by a common expanding base. In conformity with the principle of equivalence of the gravitation and inertia, the material system, expanded with acceleration, is able to move by the curved trajectory also, for example, when the external impulse on it had some angle in relation to its initial speed's direction.

According to this explanation, the mutual movement of the Earth and the Moon, for example, shows that they have formed a common body in the beginning. Each of them "remembers" the location of the general center after their division and now moves around it. This example gives an opportunity to interpret heaven's mechanics, without participation of the assumed, far-acting forces (or under the influence of the curved space-time), but on the bases of expansion and inertial phenomena only. Such opportunity ensues from the above-discussed principle of full equivalence of the gravitation and inertia. The absence of "gravitational wave" also becomes obvious from this interpretation, as the *"gravitational wave" does not at all conform to the principle of identity of gravitation and inertia.*

It is not difficult to guess that not only the former concepts of "rest" and "rectilinear movement," but also the concept of a "direct line" depreciate and become false in the expanding real world. The geometry in the expanding world may be corresponding to the *Euclid's geometry* (designed for the static world) approximately only, in the limited areas of space, where the density of substance is insignificant too. The situation for the observer of the real world is possible to imagine by an example of one architect who tries to draw the correct geometrical figures on the one elastic cloth pulled continuously from all of sides. It is possible to conclude that if our architect works quickly enough and the stretching of a cloth occurs not so strongly, the drawn figures will be disfigured insignificantly only and can be imperceptible.

The speed of action of our architect corresponds to a maximal speed of measuring that is possible to us. It's the light's speed.

The expanding cloth can serve as a two-dimensional analogue to an expanding material world. Thus, the geometry of our world will be differing from *Euclid's geometry, which is constructed for the static world with the assumption of the opportunity of instant measure.* The correct laws of new geometry for our real world will be described by more complex equations than before. The physical values, such as masses or density of substance, the gravitational constant, and the speed of the light should be considered in new geometry as the arguments together with the geometrical values. Thus, the offered concept of the expanding world enables also distinct interpretation of sense to a formal concept of GTR about the statement to be *"curved Lorentz's"* (or *"Reeman's"*) of the geometry of our world, being one of the difficultly perceived questions in this theory. The physical sense of a *"geodetic line"* in the "curved space-time," by which the bodies have moved free of influences, becomes clearly in the context of the resulted explanations. The "geodetic lines" are looking to us in the real world as the orbits of planets. In the nonexpanding mental world, these will be looking as the spirally expanded curves, by which the cosmic objects have moved by keeping their proportional, coordinated locations with all of the materials.

The one important conclusion from the concept of expansion comes out also concerning the *law of preservation of energy*. It is impossible to detect the transformation of energy from one kind to another for the separately taken body in the expanding real world. The transformation of a potential energy contained in a matter into kinetic energy of expansion will be impossible to observe in that case only if all used means of measuring remain proportional during expansion, in coordination with all materials. The marked condition can be preserved if the base of a matter is the unique kind of a physical reality that has been marked above. The obvious formulations about the uniqueness of the matter's base aren't in the GTR. However, Einstein's persistent attempts to the creation of the *general field theory* for many years show to us that he undoubtedly had deliberated the deep meaning and inevitability of this principle.

It will be very pertinent to have some stop in this connection on the alternative theory of Indian astrophysicist *Jayant Narlikar*, on a so-called *scale-invariant theory of gravitation* [L-7]. The principle of unity of a fundamental substance is distinctly expressed in the mentioned theory, unfortunately, not so popular among experts as the GTR. The essence of gravitation as the phenomenon caused by dynamically changed condition of a matter, actually, formally is formulated in the mentioned theory. The brief representation of that consists in the following.

All kinds of physical values and units of measure become possible to express through one only, considering the same as the base unit, issued from the fact of existence of the *world's universal Planck's constant* and from quantum theory— for example, by the certain length or mass. The selected unique base unit in that case may be considered as the variable also, because of described possibility. All of the physical laws and phenomenon also, known to us, in that case will remain unchangeable. The author of the mentioned theory had proved that *it's become possible the completely quantitative description of all the gravitational phenomena with an assumption of a variable condition of the accepted basic physical value.*

The *variable mass* is accepted to be the basic physical value in the mentioned theory, the origin and variation of that by the author of theory connects with the cosmic substance, in conformity to *Mach's principle*. Such explanation is unacceptable to us, as it assumes the existence of a far action and, thus, a new kind of physical reality. However, the other conclusion is important for us from the mentioned theory. From the described brief acquaintance of this theory, it becomes obvious that, actually, it gives the quantitative acknowledgement of an expanding world. It shows that:

a) *The base of all materials is the unique kind of a physical reality.*
b) *The opportunity to explain the gravitational phenomenon as consequence of a dynamically changing condition of substance.*

The rightness of the above two allegations is possible to demonstrate by the following judgments.

If the several independent physical essences are in the base of a matter's formation, then the natural constants same as the Planck's constant will be shown in several too! Thus, the universality and uniqueness of this constant is pointing to the uniqueness of a physical essence serving as the base to formation for all of the material objects, existing in diverse forms.

It is possible to tell about the second point only that the variable mass (or length in conformity with quantum representation) completely is coordinated to the put-forward concept of expansion and that may be considered as an independent acknowledgement to its correctness.

We shall stop now on the illustration of a question: *how will it look the matter's expansion process in the mental, not expanding system of readout?* The elementary quantitative reasons show that the expansion of a material world will take place by exponential laws generally. Thus, the speed of removal of a certain point, in relation to center of expansion, will be continuously increasing with time and aspiring to a light's speed. The acceleration of expansion (gravitational acceleration) will be continuously decreasing by aspiring to zero. The course of time (average frequency of a regularly repeating event) also will aspire to zero (or the interval of time will aspire to an infinity).

In the not expanded mental system of readout, the energy of matter continuously turns into kinetic energy of movement. However, described phenomena will remain undetectable for the observer of a real world. The physical laws and constants for him will remain unchangeable. Some of secondary phenomena, familiar as "gravitational," will be shown only. Thus, the gravitational forces (weight) for the observer of a real world are the passive forces. Meanwhile, actually (those not given to us to observe), these forces are active and carrying continuous transformation of energy.

It has been mentioned above about Mach's principle and about its unacceptability to us. Therefore, silently passing such intriguing question without representation of the appropriate reasoning to what's told above, it would be unjustified in this book. The summary of the essence of the question can be presented in the base of known experiment with the *pendulum of Foucoult*. It has been revealed experimentally that the plane of fluctuation of a pendulum slowly turns around in its vertical axis in dependence on the geographical breadth.

The time of a full circle of a pendulum is equal to one day on the Earth's poles, whereas on the equator it does not turn in general, for example. We know that actually it has been rotating not the pendulum, but the Earth on its axis. The oscillating pendant pendulum shows only the remarkable property to keep the direction of its oscillation aftermath of which the daily rotation of the Earth becomes possible to detect. All of the complexities here consist the following question: To what relation the position of a pendulum is to be considered as not rotated? The astronomical supervisions show that the plane of a pendulum does not rotate in relation to the remote stars. In that reason, Mach made a corresponding conclusion about the existence of an unknown influence between particles of matter and general cosmic substance. In the author's opinion, the described conclusion is not true in its root, and why is it so? The question— towards what relation to consider the rotation—it has two aspects, from mathematical and physical points of view. Actually, it is impossible to solve the question in frameworks of generalized abstract concepts of mathematics, without including external basic systems of comparison. However, the decision of a problem is not so difficult with realistic approach, if taken into consideration that in real world we always deal with the material objects and with those or others of its physical properties. The observer really will have no possibility to define what rotates whether or not his system of readout by geometrical measuring is inside of that only. Nevertheless, if you allow him to take advantage of physical experiments too, so very soon he can define the condition of his system of readout without reflecting external objects and somewhat of an interaction with them. The one small disk he can establish in the center of his system of readout for this purpose, for example. The observer may reach an absence of phenomena on the experimental (small) disk earlier taking place, by rotating that on this or opposite directions, choosing the speed of its rotation also. The centrifugal forces, forces of *Coriolis*, the concaves on the liquid's surfaces in the cups, and the rotation of a pendulum's plane are possible to neutralize this way.

As the entire group of described phenomena is possible to neutralize completely on a rotating small disk, hence, the observer can assert that all these phenomena have been caused by the rotation of its own system of readout in the opposite direction, in relation to a small experimental disk. Thus, it is possible to say that an attribute of rotation of a physical object is the difference of speeds and accelerations by ratios and by directions in its different points. However, the external systems of readout are necessary to the establishment of the evenly rectilinear movement, against rotation, which is possible to be established by physical experiments inside the system. Thus, it is possible to say as the attribute of nonaccelerated movement of the physical object, the absence of the acceleration in all of its points may be considered.

Thus, the fact that oscillation's plane of a pendant pendulum is unchangeable in relation to the far stars shows only the nonrotation of the pendulum's plane and for the cosmic substance also, but that isn't proof of the presence of an interaction between them.

The difference between revolving and not revolving objects is easier to establish from energetic point of view.

For example, the fragment, previously connected with the center of the rotated hard disk by the one spring, will lead to tension of that after separation of a fragment from the disk that's impossible to do in case of nonrotated disk. We notice that the above-described experiment is possible to make in a system of readout joined with the examined body, without the use of the idea about far stars or about cosmic matter. Thus, we can assert:

The certain kinetic energy is connected with rotated bodies, which may be exposed in the systems of readout joining them.

Thus, we always have opportunity to detect—rotated or not the physical bodies are by the shown criteria—without using the external systems of readout.

It is necessary to mark that matter is about the physical reality—energy, which is impossible to neutralize by choosing the systems of readout, by means of mathematical operations only. The above explanation is showing clearly the incorrectness of a Mach's principle, in author's viewpoint.

Now we will continue our reasoning that should be already comprehensive for the readers. Our reader knows now that the direct observation of any change with the substance are impossible because of preservation of a proportion for all kinds of material objects during its continuous expansion. However, some secondary effects are arising aftermath universal expansion because of the limitation of light's speed, at the presence of the expansion's speed and acceleration that is possible for detection.

We'll assume now that the observer's system of readout is connected with the center of a material sphere (M), and he measures the length of radius (R) (see fig. 2).

The time necessary for a light to pass a measuring distance will be R/C.

This time will increase a little bit during measuring because of expansion's speed:

$$\Delta t = R/(C - V_R) - R/C = R V_R / (C^2 - V_R C) \qquad (19)$$

The measured distance will be increased after increasing the measured time by a relative value:

$$k_V = \Delta R / R = \Delta t \, V_R / R = V_R^2 / (C^2 - V_R C) \qquad (20)$$

This expression is possible to transform to the following rows:

$$k_V = V_R^2 / (C^2 - V_R C) = V_R^2 / C^2 (1 - V_R / C) =$$

$$V_R^2 / C^2 + V_R^3 / C^3 + V_R^4 / C^4 + \ldots \qquad (21)$$

To define V_R we can use the judgments below:

The acceleration of a point allocated on the surface of the sphere in the initial moment of measuring is defined by the parity (18):

$$g = GM / R^2 \qquad (22)$$

Issued above, the instant value of expansion's speed is necessary to equal to a speed of a movement with acceleration g on the end of a way R.

In conformity with what's said above, we will get this:

$$V_R = \sqrt{2gR} = \sqrt{2GM/R} \qquad (23)$$

That value of speed in physics is named as a *"second cosmic speed"* of a material body, having the following interpretation:

If such speed will transfer to a trial body on the radial direction, then it will go far from gravitation object on the unlimited distance and will be free from its "gravitation field."

The certain kinetic energy of a trial body is corresponding to that speed (23). Thus, from the viewpoint of the expansion's concept, the same quantity of energy (speed) needs to transfer to the trial body, it being fully free and always far from expanded material objects, during the expansion process. By putting this ratio of speed in (21), we will get this expression:

$$k_v = 2 GM / C^2 R + (2 GM / C^2 R)^{3/2} + 4G^2 M^2 / C^4 R^2 + \ldots \quad (24)$$

Thus, the coefficient k_v has correcting meaning, taking into consideration the mistakes in the process of measuring because expansion's speed and limitation of light's speed.

Now, for comparison, we will look at the solution of *Shwarthshild,* deduced by him from the equations of GTR: [L-8]

$$e^v \approx 1 - r_g / R \approx 1 - 2G M / C^2 R \qquad (25)$$

The coefficient e^v, in conformity with terminology of GTR, is characterizing the "curvature of space-time," and it defines the values for some of the "gravitation effects."

The expression $r_g = 2GM/C^2$ is named in GTR as a *"gravitation radius"* for the studying object.

We can see that the correcting member in this expression corresponds with the first member in the parity (24) by its value, which gives possibility for more exact calculations by counting additional members too. Thus, we can say that the parity (24) theoretically is more correct than Shwarthshild's mentioned solution.

Now we will define the effect of the expansion's acceleration in our measurements too. The influence of the acceleration (or the decrease of a time's course) will be shown also during the above-described measuring. To define the value of this effect, we can do following reasoning:

The speed of expansion of a material sphere will be increased by some small value during the time of passage of the light's signal of the way R:

$$\Delta V_g = g R / C \qquad (26)$$

The following important circumstance is necessary to mark beforehand. Paying attention on (21), it is possible to notice the absence of a first step of the parity (V_R/C) in that row. It follows from the accepted reasoning, in view of the initial condition about the impossibility to detect the expansion's speed by the direct supervisions. The direct supervisions are the consequences of change in the expansion's speed (consequences of the acceleration). It is possible in the real world (by the example of the phenomenon of free falling). We will get an expression for the factor of acceleration, in view of the told and by the use of the previous procedure:

$$k_g = \Delta V_g / C + (\Delta V_g)^2 / C^2 + (\Delta V_g)^3 / C^3 + \dots =$$

$$GM / C^2 R + G^2 M^2 / C^4 R^2 + G^3 M^3 / C^6 R^3 + \dots \qquad (27)$$

We have supposed during our deducing of the factors k_v and k_g that the speed and acceleration have acted in our measuring independently one from the other. Thus, the same, we have admitted some discrepancies with reality. In view of the told, we are obliged to count new factors in correcting this omission. In particular, it is possible to deduce a new factor considering the action of the acceleration during minor time, increased because of the action of the expansion's speed. We can get a new factor issuing the above-said and taking into consideration the parity (19):

$$k_{gv} = \Delta t g / C = R V_R g / C \left(C^2 - V_R C \right) \approx \sqrt{2} (GM/R)^{3/2} / C^3 \quad (28)$$

It is clear that deduction of new factors can be continued as much as it will be necessary. But taking into consideration the fast reduction of their values (with assumption of the small values of speeds in relation to speed of light) at the decisions for specific problems, we can be satisfied by the first members of factors k_v and k_g only, ignoring others.

$$k_g \approx 0.5\, k_v \approx GM\,/\,C^2 R \qquad\qquad (29)$$

From the presented explanation of the gravitation, it becomes clear (for experts mostly) the possibility to do our calculations in a generalized case too, by using the three-coordinate systems of readout and excluding the approximation of physical values. It is possible also to apply the more generalized *covariant description* of a movement, thus, to reach to the certain equations, which will be equal with the equations of GTR with value and meaning. Nevertheless, in calculations of some gravitational phenomena, the use of the mentioned factors only (**29**) becomes enough.

EXPLANATIONS OF THE GRAVITATIONAL PHENOMENA FROM THE VIEWPOINT OF EXPANSION

Here are presented the same decisions from Einstein's theory deduced logically and by simple mathematics only

We shall present now the interpretations and decisions of some "gravitational effects" from the position of the expansion's concept. The use of deduced factors (k_v and k_g) can be various in the decisions in problems, depending on the initial conditions and from demanded answers. It is necessary to remember always that *all of the gravitational effects are caused by the speed and acceleration of matter's expansion and by finiteness of light's speed.*

We assume the light's signal cross a way l << R from surface of the material body, M, by the radial direction (see fig. 2). *We need to define the change of light's frequency in the "gravitational field."*

The change of frequency of light's signal is caused by Doppler's effect, according to the concept of expansion that is not difficult to define from the initial conditions of the problem. The own expansion of a way—(*l*) and light's wavy length occurs proportionally in conformity with initial assumption. Thus, the change of light's frequency will be conditioned by acceleration's factor only. To counting mentioned effect it's necessary taking into consideration that the value of acceleration is changeable on the way—(*l*). The change of light's frequency in the small time necessary to passing the elementary distance (**dR**) will be defined by the following expression:

$$\mathrm{d}f/f_0 \approx \mathrm{d}V/C \approx g_R \mathrm{d}R / C^2 = GM\mathrm{d}R / C^2 R^2$$

The summary effect of frequency's change we will define by integration of above equation:

$$\Delta f / f_0 \approx (GM / C^2) \int_{R}^{R+l} dR / R^2 = - GM / C^2 R_0 (1 / R_0 + l) = - k_g (1 / l + R_0) \quad (30)$$

In case of a condition $l << R$ from this parity it is possible to accept:

$$\Delta v / v \approx \pm gl / C^2 = \pm GM l / C^2 R^2 \approx \pm k_g l / R \qquad (31)$$

It's easy to understand that the sign of change will depend on the direction of light's way (the frequency will decrease when the light's ray goes up, and on the opposite side, it will increase). We shall note that the described effect was defined in GTR, declaring it as "influences of the gravitational field on a light," according to the following expression: [L-8]

$$\Delta v / v = \Delta \varphi / C^2 = GM (1/R - 1/R+l)/ C^2 \qquad (32)$$

($\Delta \varphi$ is interpreted as a *"difference of potentials in gravitational field"* in conformity with the accepted terminology.) This expression coordinates with (31) by its quantitative value. It is necessary to remind that the similar gravitational effects are very insignificant and difficult to detect in terrestrial scales. Nevertheless, the described experiment was made already with the use of, known in physics, *Mossbauer effect. Shapiro* has confirmed it with the others in satisfying accuracy for the small distance ($l << R$). Later *NASA* tested this consequence of GTR for the general case by using the frequency sources based on satellite (see *Gravity Probe-A* experiment).

We shall examine now the definition of curvature for the light's way near a massive body.

We shall assume that the light's beam, radiating from the remote star (S), has passed closely the massive body (M), on a distance (R_0), from its center and reaches to the observer allocated in a point (N) (see fig. 3).

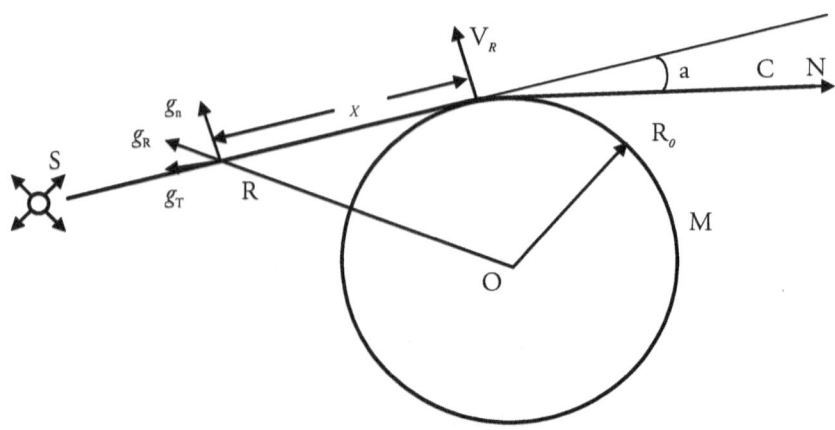

Fig. 3. The curvature of light's way near the massive body

The light's trace appears curved for the observer of the expanding world under actions of two factors—by the consequences of the expansion's speed and acceleration:

$$a = a_V + a_g \tag{33}$$

Where: a is the observable angular deviation of light's way from the rectilinear. The a_V and a_g are shares of deviation, corresponding to the factors of speed and acceleration of an expansion. From initial conditions of a problem and with consideration of the small measuring values, it is possible to define at once:

$$a_V \approx tg\, a_V = k_v \approx 2\, GM\,/\,C^2 R_o \tag{34}$$

To define a part of the acceleration in the effect, first, it is necessary to consider that the acceleration of expansion is not constantly on the all-lights way, but it changes according to parity (18). Secondly, we should consider that the effect of curvature participates perpendicularly to a way of light, the component of acceleration only. The value of a perpendicular component of the acceleration in a point corresponded to a distance x on the way of light will be defined:

$$g_n = g_R\, R_o\,/\,R = (\,GM\,/\,R^2)(\,R_o\,/\,R) = GM\, R_o\,/\,R^3 =$$

$$GM\, R_o\,/\,(x^2 + R^2_o)^{3/2} \tag{35}$$

To define the result of action of the acceleration for all the way, we need mentally to divide that into small elementary sites, for each of which g_n is possible to consider as constant then, by summation of separate elementary actions of the accelerations, define the summary action. The described procedure is named in mathematics as a method of differential calculus. The speed of expansion perpendicular to a direction will increase a little bit because of the action of a perpendicular component of the acceleration, during minor time of passage of light in the elementary way dx:

$$dV_n = (dx/C)\, g_n = GM\, R_0\, (dx/C)/(x^2 + R^2_0)^{3/2} \qquad (36)$$

In view of a great distance of light's way in comparison with the sizes of the material body, at calculations, it is possible to consider it as unlimited. (Thus, it is possible to accept that x varies in an interval - ∞, + ∞.) The summary value of speed change will be defined by the integration of the above expression:

$$\Delta V_n = GMR_0 \int_{-\infty}^{+\infty} dx/C\, (x^2 + R^2_0)^{3/2} = 2\, GM/C\, R_0 \qquad (37)$$

This value of speed corresponds to the movement of the observer's system on a perpendicular direction in comparison to a light's way. It is the cause of an observable angular change of light's direction in the observer's system corresponding to the factor of acceleration:

$$a_g \approx tg\, a_g = \Delta V_n/C = 2\, GM/C^2 R_0 \approx 2\, k_g \qquad (38)$$

Thus, the total value of a curvature for the light's way will correspond to the expression:

$$a = a_V + a_g \approx k_v + 2k_g \approx 2k_v \approx 4\, GM/C^2 R_0 \qquad (39)$$

The value of this effect also completely coordinates with consequences of GTR and for the first time has been confirmed by *Eddington* at a supervision of a solar eclipse in 1919.

The Displacement of the Orbits of the Planets

The examining problem concerns a share of a known displacement of the orbit of the planet Mercury in the rather small value of a century. The explanation of that effect was unclear for a long time within the frame of Newton's gravitation theory. The one conformable explanation is given by GTR to the angular displacement of the orbits of planets that has appreciably contributed to the trust to that with time. It is necessary to mark that the effect is extremely small, and it's much difficult to observe in relation to the other planets of a solar system. The decision of this problem from the position of general expansion's concept does not differ from others in principle and does not represent somewhat a difficulty. In the described effect, the acting factors are speed and the acceleration of general expansion, already familiar to us. The causal explanation of the observable effect is the following: The orbit of the planets expands too, according to the general expansion of a material world. The slowly rotation of the orbit's position becomes visible as a consequence of that. The observable change of the orbital length, in conformity with expansion's concept, will be defined by the following parity:

$$\Delta l \approx l \left(k_v + k_g \right) \approx 2\pi R \left(k_v + k_g \right) \qquad (40)$$

Where: R is the orbital radius.

Such expansion of orbital length corresponds to its angular displacement "forward," in a direction of a movement, conforming its value for the one full rotation cycle, to the expression below:

$$\Delta \varphi \approx tg\varphi \approx \Delta l / R \approx 2\pi \left(k_v + k_g \right) \approx 6 \pi GM / C^2 R \qquad (41)$$

Where: M is the mass of a central star.

It is clear that such change of the orbit's position will be a subject to supervision in that case only if it has the certain elliptic form. Considering the elliptic form of the orbit in a place of radius, it's accepted its average value defined by geometrical calculations. In view of what's told, the final expression becomes

$$\Delta \varphi \approx 6 \pi GM / C^2 a (1 - e) \qquad (42)$$

Where: a is the half of a big axis, and e is the elliptic parameter of the orbit. The expression (42) represents the same decision from GTR.

As we can see, the solutions for the all effects are interconnected by the common causal base and by its values, correlating each to others. It's necessary to add here one important notice only. In case the gravitation effects are caused with the expansion of the observer's system of readout and the light's signal is directed opposite of expansion, the observer will see the events as accelerated. It's correlating with the increase of light's frequency (blue shift). By this reason, the observer will see the angular displacement of the orbits on a forward direction, for example. This phenomenon is named in GTR as *"acceleration of time in a gravitation field."*

By the same cause, in the gravitation effects caused by the general expansion, by values more than local effects in the observer's system of readout (or when the red shift of light is more than the blue shift), the observer will see the same events as slowed down. Thus, in case of possibility to observing the orbital movement on the corresponding far distances, the orbit's displacement will seem to be on the opposite direction.

The "Delay" of Light's Signal in the "Gravitational Field"

From view of the expansion's concept, the speed of light does not "slow down under influence of a gravitational field" as it has commented in GTR, but its way has increased in consequence of general expansion. On the basis of the told, the additional small time for the light's signal at his passage of elementary way (dR) on a radial direction (see fig. 2), we can define by means of the deduced factors:

$$dt_A = (k_v + k_g)\, dR / C \approx 3\, GM\, dR / C^3 R \qquad (43)$$

The factors k_v and k_g are variables in the above parity, depending on distance (R). To define the total value of the signal's delay on the distance (L) from center of a material body (M), it is necessary integrating this equation for the stipulated distance:

$$\Delta t \approx (3\, GM / C^3) \int_{R_0}^{l} dR / R = (3\, GM / C^3)\, ln(L/R_0) \qquad (44)$$

The similar experiments of measuring the time of delay for the light's signals (or to radio signals) also has been made with time and was proceeded to make now, results of which coordinate both with conclusions of GTR and with a parity (44).

Thus, above has been shown the opportunity to deducing the same effects following from GTR and having the experimental confirmations, by means of comprehensively logical reasoning and elementary quantitative calculations only.

We will talk now about conclusions differed (it will more right to say, they ostensibly are different) from the expected consequences of GTR, caused by incorrectness of nowadays-accepted comments of gravitational phenomena (or by the absence of that).

It's already described in the fourth chapter the one idea of the possible experiment on acknowledgement or denial of the validity of a restriction of "localness" in the principle of equivalence of the gravitation with the inertia. As desired, it's possible to find other approaches and to offer other experiments on this direction. We had mentioned earlier another different consequence following from GTR and from the expansion's concept. What's told refers to the "gravitational wave" that's not yet detected despite the long-term attempts and on reaching unimaginably exact measuring. The existence of a "gravitational wave" is denied unequivocally from the position of the expansion's concept. Thus, the detection of it would mean the full breakdown for the suggested explanation of gravitation. On it, we can welcome only the resoluteness of experimenters on the continuation of the experiments on a task to finally proving (or to finally denying) the existence of a "gravitational wave." The speech concerns the series of grandiose experiments designed on the present time by purpose of detection of the "gravitational waves," with the use of the laser interferometers (see projects *LIGO* and *LISA*).

The following lawful question may be arisen here with the attentive readers: why are the consequences from GTR and from the concept of expansion concerning "gravitational waves" differ, whereas above it is shown the quantitative equality of them? The author's explanation to this logical question is the following. The expansion's process is presented in GTR as a property of the hypothetical "space-time" (or of a gravitational field). Thus, the "indignant condition" of it may exist also formally, having its quantitative description but nothing common with reality.

The other direction of the experiments, the expected results of which are opposite from the points of view of a present interpretation of GTR and from the concept of the expansion, is concerning experiments of direct detection on the properties of "space-time." In particular, in the writ of these pages, by *Stanford's University* together with *NASA,* is carried the unique experiment of detection of "the rotation of a space-time, involved under gravitational field of a rotating Earth" (see project *Gravity Probe—B*). The theoretical substantiation for this experiment considers the so-called effect of *Lenze-Tirring* [L-8] under influence of a *"nondiagonal components of a metric tensor,"* which is out of the Shwartzshild's solution, according to accepted terminology. From the viewpoint of expansion, the named component by its structure and by its value corresponds to the second member in the row of k_V in (24) and to the first—in the expression—k_{gV} in (28), deduced above. What's told means these members can have only the small correction of meaning in the already-known phenomena. Thus, these can't cause the occurrence of new kinds of gravitation effects. It follows to expect the negative result in the described experiment, according to expansion's concept. Based on the previous material, it's possible to give the following precise definition for prediction of the results of the possible new experiments, connected with gravitation:

Theoretically calculated, all of the effects based on the concepts of identity, the gravitation and inertia can be proven experimentally. The gravitational experiments, which are beyond of principle of identity, can lead only to negative results.

The Identity of the Gravitation and the Inertia from Energetic Point of View

We shall examine now the decision of the problem showing full equivalence of the gravitation and inertia from energetic point of view. The concept of "gravitational energy of communication for the material body" is introduced in existing works, issuing from the accepted field interpretation of gravitation. The value of that defined by the following mental experiment—the body disassembles on many smallest pieces and removes them on the infinite distances each from another. The demanded quantity of energy to perform such operation accepted and named as the full potential gravitation energy of the body. The value of potential energy, based on the field representation and on described mental experiment, is defined by expression: [L-7]

$$E_g = 3GM^2 / 5R_0 \qquad (45)$$

From expansion's viewpoint, the energy named "gravitational" is the kinetic energy, caused by the expanded condition of the matter. Based on what's told, we will define the value of a kinetic energy connected with expansion for the one homogeneous, spherical solid body (see fig. 4).

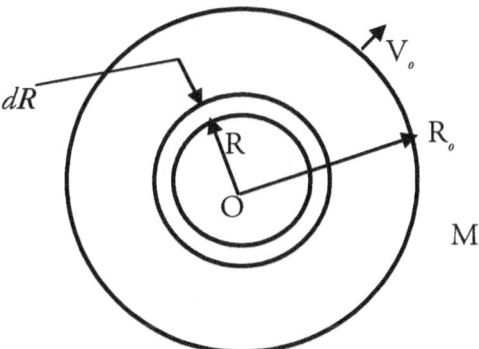

Fig. 4. The energy of an extending body

From fig. 4 and from the initial concept of expansion, it is possible to write:

$$V_0 = \sqrt{2\,GM/R_0} \qquad (46)$$

For the elementary quantity of a matter, contained in a volume between spherical surfaces R and $R + dR$, the kinetic energy of expansion will be defined:

$$w = dM \, V_R^2 / 2 \qquad (47)$$

The mass contained in elementary volume will be defined:

$$dM = 4\pi \, R^2 \rho \, dR = 3M \, R^2 dR / R_0^3 \qquad (48)$$

The speed of radial movement of the elementary mass will be defined:

$$V_R = V_0 R / R_0 = R\sqrt{2\,GM/R_0} \Big/ R_0 = R\sqrt{2\,GM/R_0^3} \qquad (49)$$

The kinetic energy of the elementary mass will be defined:

$$d \, W_k = 3GM^2 R^4 dR / R_0^6 \qquad (50)$$

By integration of the received expression for the interval $(0 - R_0)$, we will get:

$$W_k = (3GM^2/ R_0^6) \int_0^{R_0} R^4 dR = 3G \, M^2/5 \, R_0 = E_g \qquad (51)$$

The equality of the resulting expression (51) with the above mentioned, deduced on the different conceptual interpretations—on gravitation and inertia, shows the actually identity of these names from energetic point of view.

It's necessary to mark the representation of the gravitational energy as the kinetic, caused by expanded condition of a matter; it removes one of important critical remarks to GTR. The matter consists in the theoretical conclusion of *dependence of the gravitational energy from choosing system of readout.* The marked dependence is difficult in conforming the field representation of gravitation; therefore, it has caused the hard disputes with time (particularly between academicians *Logunov* and *Ginzbourg*). Everything becomes in their places in conformity with the concept of expansion because the dependence of a kinetic energy from the system of readout, where it is defined, it is elementary obviously and not requires especial explanation.

THE EXPANSION AS A FUNDAMENTAL
PRINCIPLE OF A MATTER
THE GRAVITATIONAL CONSTANT

The mystery hasn't been in the supposed essences but it was in the initial bricks . . .

From the logical viewpoint, it will be right to present this chapter after detailed representation of the principles and the structure for that "unique fundamental reality," by properties of the existence and expansion of matter that are conditioned; that is, it contains in a plan the author's second book devoted to marked problems. However, by coming into situation of a reader probably wishing already to reach to the logical end of the gravitation's history, below in form of short theses, we shall present the author's approach to the explanation of the deep essence of the matter and to the causal interpretation of its expansion. From previous pages, the reader already had known about some circumstances that force to think about the uniqueness of the physical reality deserving as the basis for the formation of all the materials. Below we will bring now some allegations, based on the already mentioned and on the other logical arguments:

1. *The electromagnetic field serves as the unique kind of physical reality forming the material objects.* It consists of the elementary indivisible "grains" that we shall name the *"quantum of a field"* with the preliminary agreement about some difference of the sense of this name from being accepted. The matter is, this name is identified with the *photon* in present time, but we shall mean under this name the next general sense. According to the author's consideration and used terminology below

2. *The photons and every possible known and unknown elementary particle also (by accepted meaning of these names) are the various displays of the electromagnetic field's quantum.*

Thus

3. *The quantum of field may be displayed in two possible physical conditions: in the form of linearly spreading group of waves—as a photon—and in the form of a localized condition of the same group of waves representing the various other elementary particles (except of the photon).*
4. *The localized quantum of the field is characterized by a geometrical constant, by the constant of fine structure, conditioned by its wavy nature (similarly to π, which characterizes the circle, sphere, etc.).*
5. *The Planck's constant and the speed of light are the natural physical constants characterizing the quantum of a field quantitatively and dynamical.*
6. *The value of energy and all kinds of parameters of the field's quantum are mutually interconnected and may be defined by only one from them—with the free parameter of its condition (the energy, wave frequency, or its length, etc., may be taken as quantum's free parameter).*

It further shows that

7. *The localized wave formations of the electromagnetic field's quantum may be especially stable in some rare conditions and unstable in general case, corresponding with the known properties of the different kinds of stable and unstable elementary particles.*

The numerous unstable elementary particles are being represented as the different transient conditions between the localized and nonlocalized field's quantum, in conformity with the described concept. Thus, by this clear reason, it will be right to look at the nonstable particles as the objects having no future in studying it. The answer of such question, on what basis and by which principle the electron is constructed, should be more important to us than the openings of the thousands miraculously new particles living a milliards of share of a second.

The above-suggested conceptual picture of matter's formation opens an opportunity to description of the electron and other stable fundamental particles also the proton, the neutron, and neutrino, with all of its experimentally known properties. Meanwhile, the offered field representation of the elementary particles isn't something that completely deletes the importance of the quantum theory or other formalistic approaches, despite of the total difference of used languages and applied methods. It will be right to say that the suggested field's representation of elementary particles may be looked as the causal interpretation of a microcosm and the phenomena occurring in it, by an analogy of the expansion's concept. Thus, it gives the logical interpretations to known facts in this area, by filling up the same formal theories, without contradiction to them, in the author's belief.

Being limited with the resulted acquaintance, we shall present now the conceptual explanation of a matter's expansion, according to the above representation of the elementary particles. For this purpose, some words about the sense of the mentioned *constant of a fine structure* are necessary to tell preliminarily. It should be known to our reader in present time the meaning or causal origin of this numeric constant, concerning the sets of phenomena in the microcosm, there isn't any explanation or even any serious hypothesis. The mentioned constant has been appearing in our formulas exclusively from experimentally getting the results only. According to author's conception

8. *The concentration of energy into localized quantum of field is characterized by the constant of a fine structure. It defines the parity of energy of the electric and magnetic static fields to a full energy of a localized quantum, corresponding to the mass of the formed elementary particle. By the constant of fine structure, same time is defined the natural uncertainty of a localized quantum of field (in accepted terminology, it is corresponding with the uncertainty of parameters of a quantum object).*

Proceeding from the resulted description, the natural expansion of a particle directly contacts the author with the uncertainty of localized quantum's parameters. Thus

9. *The harmonious and self-coordinated expansion of the universe occurs aftermath a continuous expansion of the electromagnetic field's quantum being the basis for all of the material objects. According to named approach, it is possible to conclude that the value of the gravitation constant should be defined (or to be set) by the constant of fine structure. As well as the set of the other properties of the elementary particles, the gravitational constant also should be expressed by the basic natural constants of the quantum of a field by Planck's constant and by speed of light.*

We will look the electron as the investigating particle, by a reason of its more simple "construction" in comparison with the other particles (about that, it will be possible to judge in the future book.) In the beginning, we shall bring the formula, according to which the size of the electron is defined, on the basis of a described wavy representation of the elementary particles:

$$r_e = \lambda_k \, (1 + a/2\pi) \, /2\pi \qquad (52)$$

Where: r_e is the average radius of distribution for the mass of the electron; λ_k is the *Compton's wavy length* for the electron; a is the *constant of fine structure*.

It is necessary to inform the reader that the marked value on a first sight does not coordinate with the **quantum-mechanical imagination of the electron** nowadays accepted. However, there is no final comprehension on this direction, and thus, there isn't any direct contradiction to this expression. According to the described assumption, the natural expansion of the electron was accepted as equal to the following:

$$V_e = aC \left(t_k / T \right) = a \lambda_k / T = a \lambda_k / sec. \tag{53}$$

Where: V_e is the speed of a natural expansion for the electron; t_k is the period of oscillation of the Compton's wave for the electron; T is the accepted unit of time ($T = 1$ *second*).

To explain the meaning of the expression (53), some deviation is necessary. As already it is shown in the fifth chapter, it's incorrect to operate with the concept of the abstract time. The time should be defined as a property of the matter and should be connected with that at descriptions of the gravitational phenomena. In the macrocosm, the course of time (the average frequency of the regularly occurring events) for the material objects, which are constructed from set of separate particles, was in direct proportion to an average density of substance or, in view of an equivalence between mass and energy to a density of an energy, by clear explanation.

10. The course of time for the separately taken quantum is identical with its frequency or, that the same, it is directly proportional to its energy.

The time unit accepted in physics (second) directly isn't connected with matter. In other words, the used time unit isn't the natural value, but it's the freely taken value. For rational and correct descriptions of the phenomena, connected with expansion of a matter and with change of its parameters, it is necessary to use the local time unit, directly defined by matter (by following this instruction, for example, earlier we had deduced the Newton's law of gravitation). With the described reason, as interval of time in this case, it's necessary to accept the period of fluctuation of the quantum, forming the electron (it corresponds to the above-stipulated condition of direct proportionality of the course of the time to the energy of a particle). The correcting factor (t_k / T) in the expression (53) is added to be taking in consideration of the "wrongness" of a time unit, used in physics. The similar compulsorily used correcting factors always appear in the physical equations subsequently of unsuccessfully (freely) selected systems of units (it is accepted to name them as the *systems factors*). In connection with the discussed question, it seems purposeful to refer again to above-mentioned Narlicar's approach about the establishment of physical units of measurements, as per the author's explanation actually means the same. In view of the parity $t_k = 1/v_e = \lambda_k / C$, we get to (53).

By identifying the described expansion of the electron with its gravitational expansion, we get to the equation:

$$V_e = \sqrt{2G_t m_e / r_e} = a\lambda_k / sec \qquad (54)$$

Where: m_e is the mass of the electron, G_t is the gravitational constant.

According to presented wavy representation of the elementary particles, in view of equivalence of mass and energy (1), it is possible to replace m_e with the expression:

$$m_e = h v_e / C^2 = h / C \lambda_k \qquad (55)$$

(This is well-known parity in the quantum theory.)

Where: h is the Planck's constant; v_e is the frequency of the Compton's wave for the electron.

In view of (54) and (55) from (52), the final expression follows, defining the theoretical value of the gravitation constant, according to the described initial assumptions:

$$G_t = a^2 (1 + a/2\pi) C\lambda^4_k / 4\pi h (sec.)^2 \qquad (56)$$

We get the following value for the gravitational constant having made calculations:

$$G_t \approx 6.66* 10^{-11} [Nm^2 / kg^2] \qquad (57)$$

We shall bring the experimentally defined value of the gravitation constant to comparison:

$$G \approx 6.67* 10^{-11} [Nm^2 / kg^2] \qquad (58)$$

The closeness of the deduced value to the experimental opened its difficulty to attribute it just to a casual coincidence, in author's opinion. Of course, by selection of physical constants and mathematical operations in some cases, it's possible that "the organization" of some desirable results (the similar technique is ordinary in modern formal methodology that has been noticed). However, to do it on somewhat logically consecutive causal basis is more difficult. Besides, the absence in the equation (56) the suspicious numerical factors, much different from unit by their values, indirectly testifies to correctness of a deduced equation. However, in the general context of the book, it will more lawfully look:

As the main argument to correctness of the suggested explanation and to achieved results, hence, to consider, the opportunities to deduce the Newton's law of attraction, to explain the gravitational phenomena and to define the gravitational constant by the single and general cause only; that is, the matter's expansion.

EPILOGUE

In the suggested brochure, the author had strived to show the deep inaccuracy of our imagination in the unchangeableness of the various objects surrounding us and composing our material world. All the kinds of material objects have dynamical formation and by their essence are in a permanently expanding condition, according to the author's conception. The gravitational phenomenon is the afterward of that. Such change of our imaginations and beliefs is a difficult problem from psychological point of view. Therefore, it's more connected with time than with studying the necessary proofs or arguments. In this sense, the offered concept of gravitation and the depicted picture of outlook, most likely, are destined to the future generation than to the official science and to the modern researchers; they have already their own firm beliefs and traditional approaches. It is possible to imagine, for example, that future physicists from school bench are mastering that the Earth is not only spherical and that it not only rotates on its axis or moves by its Solar orbit. Nevertheless, that also continuously increases by size, common with every possible material object because such expansion is peculiar to a fundamental principle of matter in general. They also study how the average frequency of regularly repeating events is continuously slowed down in our world, as a consequence of continuous reduction of the density of matter, etc. All of the above-mentioned allegations will be difficult or easy to master equally to those who listen first time about that, because all of those events are in direct contradiction to our perceptions too. For the physicist or for the philosopher of new generation, grown and formed in such spirit, the understanding of all unnatural representations about absolutely constant sizes and times will be so simple too, as now for us to realize the wrongness of the concepts of the absolute movements and of the absolute directions. However, the same has been unimaginable and unacceptable to our ancestors in his time too. Afterward of that, the known ancient scientific hypotheses have been composed about three elephants or whales, keeping the Earth, or about Atlantes keeping the sky with the stars, etc., to protect only their own intuitive perceptions by all means. Our situation is very similar to that now. The theorists are using various approaches to explain the gravitational phenomena, simultaneously holding their intuitively established beliefs, whereas such way often leads toward deadlock that we can find many times in the history of science; and we see it now too.

Unfortunately, by nature, the ability is given to us by the subjective perception of the insignificant part only from the complete picture of a material world surrounding us. On that reason, in the way to analytical knowledge of a reality, it's often enough when it becomes necessary for us to go counter our own intuition, sometimes muffling that, preferring the facts and calculations. Our reader now is in such unenviable position. It's necessary temporarily to forget about intuition and direct sensations of a reality to overcoming it. However, the author is hopeful with the existing lessons and acquaintance with the similar problems. We see from history that though with difficulty and often accompanied by the dramatic events new ideas sometimes found the places in science and in due course were fixed, especially, when they contained also a kernel of truth. The author deeply trusts in the truthfulness of his own conclusion and by this has taken such unpromising intention, to offer to the public his unconventional explanation of gravitation, while in the scientific world of various approaches and theories to the given problem in a sufficient set, including the well recognized and preferable. Of how much the offered decision of the old riddle of gravitation and the author's approach will be comprehensible and acceptable, it will show in time.

Author: Kirakosyan Gevorge Shawarsh   Dubai   30-05-2005

BIBLIOGRAPHY

(The used literature in a Russian language)

1. *V.B. Braginskiy*   — "Experimentalnaya proverka teoryy
   "Znanie" 1977   otnositelnocty"

2. *A. Einstein. L. Infeld.*   — "Evolutia physiki"
   "Mir" 1968

3. *A. Einstein*   — "Sushnost teoriy otnositelnocty"
   "Mir" 1955

4. *V.L. Ginzburg*   — "O teoryy otnositelnosty"
   "Nauka" 1979

5. *A. Karrelly. K. Meller & others* — Astrophisika, quanty I teoriya otnositelnosty"
   "Mir" 1989

6. *A.A. Logunov*   — "Lekcyy po teoryy otnositelnocty"
   "MGU" 1984

7. *J. Narlicar*   — "Neistovaya Vselennaya"
   "Mir" 1985

8. *G. Sahakyan*   — "Prostranstvo—vremya i gravitacia"
   "Yerevan University" 1985

9. *E. F. Tayloor. J. A. Willer & others* — "Physica prostranstva—vremeny"
   "Mir" 1969

10. *Y.B. Zeldovich. I. D. Novikov* — "Teoriya tyagoteniya I evolutiya zvezd"
    "Nauka" 1971

www.ingramcontent.com/pod-product-compliance
Lightning Source LLC
Chambersburg PA
CBHW021859170526
45157CB00005B/1880